Computer Aided Design Guide for Architecture, Engineering and Construction

Recent years have seen major changes in the approach to computer aided design (CAD) in the architectural, engineering and construction (AEC) sector. CAD is increasingly becoming a standard design tool, facilitating lower development costs and a reduced design cycle. Not only does it allow a designer to model designs in 2 and 3 dimensions, but also to model other dimensions, such as time and cost, into designs.

This *Computer Aided Design Guide for Architecture, Engineering and Construction* provides an introductory explanation of all common CAD terms and tools used in the AEC sector. It describes each approach to CAD with detailed analysis and practical examples. Analysis is provided of the strength and weaknesses of each application for all members of the project team, followed by review questions and further tasks.

Coverage includes:

- 2D CAD
- 3D CAD
- 4D CAD
- Building Information Modelling
- nD modelling

With practical examples and step-by-step guides, this book is essential reading for students of design and construction, from undergraduate level onwards.

Ghassan Aouad is the Pro-Vice-Chancellor for Research and Innovation at the University of Salford. He is also Co-Director of the £5m EPSRC-funded Salford Centre for Research & Innovation in the Built & Human Environment, a visiting professor at Universiti Teknologi, Malaysia (UTM), and Fellow of the CIOB. Professor Aouad has spent the last 20 years teaching and researching subjects related to information modelling and visualisation, nD simulation and process mapping.

Song Wu is the Programme Director for BSc (Hons) in Quantity Surveying at the University of Salford. His research interests include product and process modelling, data modelling and computer simulation. In 2010, Dr Wu was awarded the UK China Fellowship for Excellence for his collaborative research with leading Chinese research institutions.

Angela Lee is the Director of Postgraduate Taught Studies and Programme Director of the BSc (Hons) Architectural Design & Technology course within the School of the Built Environment, University of Salford. Her research and teaching centres on design management, process management, performance measurement and nD modelling.

Timothy Onyenobi (BSc Hons, MSc, PhD, MNIA, ICIOB, FInstCPD) is a Chartered Architect (Nigeria) and a Research Fellow with SCRI Sobe, University of Salford. He specialises in CAD and BIM and has been involved in numerous architectural projects in Nigeria and the UK.

Computer Aided Design Guide for Architecture, Engineering and Construction

Ghassan Aouad, Song Wu,
Angela Lee and Timothy Onyenobi

Spon Press

an imprint of Taylor & Francis

LONDON AND NEW YORK

First published 2012
by SPON Press
2 Park Square, Milton Park, Abingdon, Oxon OX14 4RN

Simultaneously published in the USA and Canada
by SPON Press
711 Third Avenue, New York, NY 10017

SPON Press is an imprint of the Taylor & Francis Group, an informa business

British Library Cataloguing in Publication Data
A catalogue record for this book is available from the British Library

Library of Congress Cataloging in Publication Data
Aouad, Ghassan.
    Computer aided design guide for architecture, engineering, and construction /
    Ghassan Aouad, Angela Lee, Song Wu, and Timothy Onyenobi.
        p. cm.
    1. Computer-aided design. 2. Architecture–Computer-aided design. I. Lee, Angela,
    Ph. D. II. Wu, Song, Ph. D. III. Onyenobi, Timothy. IV. Title.
    TA174.A58 2012
    604.2–dc23   2011023980

ISBN: 978–0–415–49505–9 (hbk)
ISBN: 978–0–415–49507–3 pbk)
ISBN: 978–0–203–87875–0 (ebk)

Typeset in Sabon
by Swales & Willis Ltd, Exeter, Devon

Printed and bound in Great Britain by
TJ International Ltd, Padstow, Cornwall

Contents

1   Introduction to CAD for the AEC/FM industry   1

   1.1 Introduction 1
   1.2 What is CAD? 1
   1.3 A brief history of CAD 3
   1.4 CAD technology 9

2   Project and product modelling   11

   2.1 Introduction 11
   2.2 Why a process model? 11
   2.3 What is a process? 16
   2.4 Approaches to process modelling 17
   2.5 Product modelling 25
   2.6 Summary 28

3   2D CAD   31

   3.1 An introduction to 2D CAD drafting 31
   3.2 The history of 2D CAD drafting 31
   3.3 2D drafting principles 36
   3.4 2D CAD practical examples 41
   3.5 Summary 48

4   3D CAD   51

   4.1 An introduction to 3D modelling 51
   4.2 3D modelling principles 55
   4.3 Creating a 3D model 60
   4.4 3D modelling practical examples 66
   4.5 Summary 69

5   **BIM (Building Information Modelling)**   71

    5.1 An introduction to BIM 71
    5.2 BIM applications within the AEC/FM industry 74
    5.3 The advantages and disadvantages of BIM 81
    5.4 BIM modelling principles 83
    5.5 Summary 91

6   **4D CAD**   93

    6.1 An introduction to 4D CAD 93
    6.2 4D CAD in practice 95
    6.3 The advantages of 4D CAD 96
    6.4 The limitations of 4D CAD 97
    6.5 The 4D CAD modelling process 97
    6.6 Summary 104

7   **nD Modelling**   107

    7.1 Introduction 107
    7.2 What is nD modelling? 109
    7.3 nD modelling research development 110
    7.4 The future of nD modelling 112

    Bibliography   117
    Index   127

Chapter 1

Introduction to CAD for the AEC/FM industry

1.1 Introduction

Why do some organisations adopt/universities teach a particular design application? Why do others use/teach different applications? What are the differences between them? Why do you use particular software applications for different purposes? Which is the right one to use?

If these questions have played on your mind, then this book is for you. Without advocating particular proprietary software packages, this book will describe the concept of computer-aided design (CAD) and help to define the purpose and scope of the colossal number of CAD systems available in the marketplace that are applicable for the architectural, engineering, construction and facilities management (AEC/FM) industry.

It was only 30 years ago that nearly all design drawings were produced by hand with pen and paper. Small changes in the design led to erasing and redrawing, whereas major changes often meant starting drawings again from scratch. An old adage commonly heard in architectural practices reflected this practice: 'never draw more in the morning than you can erase in the afternoon.'

CAD fundamentally changed this process and has enabled more complex designs. This chapter introduces the various parameters of CAD as a backdrop for the chapters that follow in this book.

1.2 What is CAD?

Definition

Computer-aided design (CAD) is the use of technology to aid the design and, particularly, the drafting of a part or product, including entire buildings. It is both a visual- (or drawing-) and symbol-based method of communication, following standard conventions to a specific technical field, such as architecture or engineering.

The process of architectural design is immensely complex. A client wanting to commission a project will traditionally approach a design expert who will try to understand their needs and propose a solution. With inexperienced clients, this dialogue may often involve verbal descriptions, and with more hands-on/experienced clients, producing cut-out imagery, photos, etc. Clients/users face a number of problems in translating their design thoughts; they usually struggle to describe their sensual, physical and cognitive ideas into words; and without prior drawing skills, they usually find it difficult to capture or understand adequately the three-dimensional nature of a building design on a flat sheet of paper. They are generally faced with the rhetorical problem of how three-dimensional space is to be designed, understood and communicated two-dimensionally, since space is not static but fluid and needs to accommodate movement (Lefebvre, 1991). To aid understanding, some designers have turned to a range of mediums to express design ideas with clients, such as Will Alsop with paintings (Evans, 1997), and Ricardo Bofill with poetry (Bofill, 2010), but more often than not, result to hand drawing and the use of CAD.

The design will then go through a series of iterations as the client/users try to understand and tailor the design to their needs (here, CAD is of benefit as the design can be quickly changed rather than being wholly redrawn by hand). The design changes as interim solutions are generated as the client/users' needs become more clearly defined through the interaction of the designer and the client/user throughout the design process.

Designers are trained to interpret and solve design problems by graphic mediation, be it through drawing, model-making or via CAD (computer-aided design; Lawson, 1994; Stern, 1977). The objective of design is the creation of form (Robbins, 1997). The purpose of graphic mediation is to give the form some shape or expression. Form is grounded in function and meaning; the legitimacy of form is based upon pro-grammatic clarity (Alexander, 1964). Thus, the designer must define the problem in order to determine its functional characteristics before expressing the form (Lee et al., 2005). The drawing/CAD/model helps achieve fitness between the brief, the context and the design (Robbins, 1997). They are representational and being forward-looking, allow for the meaningful ordering of things in the environment (Lawson, 1990). The form which grows from this fit is judged by both its functional efficiency and its formal qualities as a design. The drawing/model is, therefore, a tool in the achievement of a best-fit solution functionally and aesthetically; a tool which is evolutionary in nature, interactive with other media (models and CAD) and open to critical examination at different stages in the process (Graves, 1977; Bowers, 1999; Schon, 1983). This design process remains fluid, and is a means to engage with clients/end-users.

Parts or products can be represented in two dimensions (2D; see Chapter 3) and/or three dimensions (3D; see Chapter 4). CAD can also be used throughout the design process and for dynamic mathematical modelling, from the strength and dynamic analysis of assemblies (often marketed as CADD – computer-aided design and drafting), to the definition of manufacturing methods and components, which is often termed computer-aided manufacturing (CAM).

CAD software assists designers and engineers in a wide variety of industries to design and manufacture products ranging from automobiles, roads, aircrafts and all types of buildings, through to digital cameras, mobile phones, clothes and computers.

It has become an essential tool for modern design, and has helped to lower product development costs and to shorten the design process.

Discussion

- What is CAD used for in your profession?
- What CAD packages are you aware of? And what are the differences between them?
- What governs your choice of CAD packages?

1.3 A brief history of CAD

First recorded CAD system

In order to understand the scope and context of CAD in the AEC/FM industry, it is important to trace its development. The first recorded graphic system was developed by the Massachusetts Institute of Technology's (MIT) Lincoln Laboratory in the mid-1950s for the United States Air Force's Semi Automatic Ground Environment air defence systems. Ivan Sutherland extended this technology for his Ph.D. thesis at MIT in 1960 to develop a program called Sketchpad, which is considered to be the first step towards CAD. It was the first recorded tool that enabled the designer to interact with the computer graphically using a light pen to draw on the computer's monitor. However, the first commercial CAM software was PRONTO, a numerical control programming tool that had been developed in 1957 by Dr Patrick Hanratty, who is most often referred to as the 'father of CAD CAM'.

Due to the high cost of early computers and to the unique mechanical engineering requirements of aircraft and automobiles, large aerospace and automotive companies were the earliest commercial users of CAD. First-generation CAD software systems were typically 2D drafting applications which were primarily intended to automate repetitive drafting chores, but which also enabled scaling (an increase or decrease in scale) and rotation of design elements.

Widespread CAD implementation

Since the 1950s, the development of CAD systems has been well documented. Table 1.1 details a number of key developments. Notably, personal computers (PCs) first appeared in the early 1980s, which made CAD applications more widespread; Autodesk, founded in 1982, launched the first CAD software for PCs – 'AutoCAD Release 1'; Bentley Systems developed and released 'MicroStation' in 1984; Micro-Control Systems developed the first 3D wire-frame CAD software, named 'CADKEY', in 1985; and, with Apple releasing the first Macintosh 128 in 1984, in the following year Diehl Graphsoft developed 'MiniCAD', which rapidly became the best-selling CAD software

Table 1.1 Key developments in CAD

Year	Development
Mid-1950s	First graphic system launched by the US Air Force's SAGE (Semi Automatic Ground Environment) air defence system – developed at Massachusetts Institute of Technology's Lincoln Laboratory.
1957	Dr Patrick J. Hanratty, known as 'the Father of CADD/CAM', developed PRONTO, the first commercial numerical control programming system.
1962	The first CAD programs used simple algorithms to display patterns of lines initially in two dimensions, and then in 3D. Early work produced by Professor Charles Eastman at Carnegie-Mellon University; the Building Description System, a library of several hundred thousand architectural elements which can be assembled and drawn on screen to form a complete design concept, was established.
1971	MCS was founded and enjoyed an enviable reputation for technological leadership in mechanical CADD/CAM software. In addition to selling products under its own name, in its early years MCS also supplied the CADD/CAM software used by such companies as McDonnell Douglas (Unigraphics), Computervision (CADDS), AUTOTROL (AD380) and Control Data (CD-2000) as the core of their own products. It is estimated that 70% of all the 3D mechanical CADD/CAM systems available today trace their roots back to MCS's original code.
1972	The earliest Intergraph (M&S Computing) terminal was designed to create and display graphic information. Composed of unaltered stock parts from various vendors, the terminals consisted of a single-screen Tektronix 4014 display terminal with an attached keyboard and an 11" x 11" 'menu' tablet that provided the operator with a selection of drawing commands.
1974	First commercial sale of an M&S system; based on a PDP central processor from Digital Equipment Corporation, it ran the first version of Intergraph's original core graphics software, the Interactive Graphics Design System (IGDS), and was used for mapping applications.
1975	Avions Marcel Dassault (AMD) purchased CADAM (Computer-Augmented Drafting and Manufacturing) software equipment licences from Lockheed, thus becoming one of the very first CADAM customers.
1977	Avions Marcel Dassault assigned its engineering team the goal of creating a three-dimensional, interactive program, the forerunner of CATIA (Computer-Aided Three-Dimensional Interactive Application). Its major advance over CADAM was that all-important third dimension. While CADAM automated the existing world of two-dimension engineering, essentially drafting and calculation with roots in descriptive plane geometry, CATIA lifted Dassault engineers into the world of 3D modelling, removing the possibility of misinterpreting two-dimensional data and generating a host of immediate benefits.
1981	Dassault Systems is created.
1982	CATIA Version 1 is announced as an add-on product for 3D design and surface modelling.
1982	Autodesk founded by 16 people in California under the initiative of John Walker whose idea was to create a CAD program for a price of $1,000 which could run on a PC. The first version of AutoCAD was based on a CAD program called MicroCAD, written in 1981 by Mike Riddle.

Table 1.1 Continued

Year	Development
1984	Hungarian physicist Gabor Bajor smuggled two Macs into his country. At the time, ownership of personal computers was illegal under Communist rule. Using Pascal, he and a teenager called Tamas Hajas worked to write a 3D CAD program for the Mac which was the beginning of the Graphsoft Company.
1985	Keith Bentley founds Bentley Systems, Inc. Originally named PseudoStation, the software developed by Bentley Systems allowed users to view IGDS drawing files without needing Intergraph's software. The next version of PseudoStation was renamed MicroStation and added the ability to edit IGDS files. After Intergraph purchased 50% of Bentley Systems, a new version of MicroStation added proprietary extensions to the IGDS and renamed it DNG.
1985	Diehl Graphsoft, Inc. is founded and the first version of MiniCAD is shipped in the same year. MiniCAD became the bestselling CAD program on the Macintosh.
1986	AutoCAD received 'The Best CAD Product' award from PC World magazine, and every year subsequently for the next 10 years.
1988	CATIA Version 3 released with AEC functionality. CATIA is ported to IBM's UNIX-based RISC System/6000 workstations and becomes application leader in the automotive industry.
1989	Autodesk buy Generic Software and Generic CADD program, making over 600 add-on applications for AutoCAD.
1990	Visio Corporation founded, producing graphics and drawings software.
1991	Microsoft develops Open GL for use with Windows NT. Open GL is an API procedural software interface for producing 3D graphics and includes approximate 120 commands to draw various primitives such as points, lines, and polygons. Open GL, developed by Silicon Graphics, is a standard for 3D colour graphics programming and rendering.
1991	First AutoCAD program for SUN platforms.
1992	Dassault Systems decides jointly with IBM to transfer the responsibility of CADAM to Dassault Systems of America, a company created in 1992 as a wholly owned subsidiary of Dassault Systems. IBM agrees to acquire a minority interest in Dassault Systems. Since then, CATIA and CADAM have become progressively unified by merging the best technological features of both systems.
1993	John Hirschtick from Computervision founds a new CAD company called SolidWorks, Inc.
1993	First AutoCAD (Release 12) for Windows platforms. It required 8 MB RAM and 34 MB Hard Drive space for complete installation. The Windows version of AutoCAD included a 36-icon toolbox, allowed multiple AutoCAD sessions, separate Render window, support for Windows GUI, DDE, and OLE, as well as Drag-and-Drop and Bird's Eye view capabilities. AutoCAD 12 for Windows was one of the most successful CAD programs ever.
1995	Unigraphics on Microsoft Windows NT debuted.
1995	CATIA-CADAM AEC Plant Solutions announced. This next-generation object-oriented plant modelling system enables powerful knowledge-based engineering capabilities that can dramatically streamline the process of plant design, construction, and operation. It brings the power of 'smart' applications to the desktop with next-generation object-oriented modelling.

Table 1.1 Continued

Year	Development
1995	Autodesk ships the first version of 3D Studio for NT platform, called 3D Studio MAX.
1995	Parametric Technology ships Pro/E version 15, the first parametric modelling CAD/CAM program and the first high-end 3D solid modelling package available on NT platforms.
1995	Dassault Systems ships ProCADAM, a shorter version of CATIA for use on NT systems.
1996	General Motors signs largest CAD/CAM contract in history, selecting Unigraphics as its single vehicle development software platform. Parasolid rapidly gains widespread penetration of the market as the de facto standard for the development of high-end, mid-range, and commercial CAD/CAM/CAE programs.
1996	CATIA-CADAM Solutions Version 4 is made available on Silicon Graphics, Hewlett Packard and Sun platforms.
1997	Autodesk ships 3D Studio MAX release 2 and a cut-down version called 3D Studio Viz.
1997	ISO 13567 – Standard for Structuring Layers in CABD – an international effort to structure layer naming for translation between various languages and vendors.
1997	Revit Technology Corporation revolutionises building design with Revit, the world's first parametric building modeller developed for the AEC industry.
1998	Dassault Systems and IBM announce a new Strategic Alliance to address the Product Development Management II (PDM II) market. Dassault Systems acquires Matra Datavision, and are appointed an IBM International Business Partner to market, sell, and support CATIA, CATweb, and ENOVIA, as well as IBM's e-business Solutions.
1998	Unigraphics Solutions becomes first CAD/CAM/CAE/PDM organisation to be awarded ISO 9001/TickIT Certification.
1998	AutoCAD Mechanical, which integrates a mechanical tool in AutoCAD 14, is launched.
1998	Autodesk Architectural Desktop, an integrated architectural solution based on AutoCAD 14, is launched.
1998	Viso Enterprise, a technical drawings and documentation program, is launched.
1999	National CAD Standard (NCS) 1.0 released – the first compendium and coordination of efforts.
1999	VectorWorks is released as a replacement for MiniCAD.
2000	Graphisoft offers a set of Web-based tools to help encourage the AEC CAD community to adopt its GDL (Geometric Description Language) as a file format. The GDL Object Web plug-in for ArchiCAD users delivers GDL objects via the Internet.
2000	IBM and Dassault Systems launch Version 5 Release 5 of CATIA, available for Windows and UNIX.
2008	Autodesk launched Revit Series 2007, one of the most popular BIM-based CAD applications.
2010	GoBIM launch the first BIM application for Apple iPhone and iPad.

Source: Adapted from Mbdesign, 2010; Cadazz, 2010; and THOCP, 2010.

for the Mac computer. Although PCs and Macs steadily increased in power throughout the 1980s, and AutoCAD™ continued to gain substantial market share in the 2D CAD software market, the general lack of processing power, and especially the poor graphics performance when compared to UNIX workstations, meant that it was not until the next decade that PCs began to have their revolutionary effect on the CAD software industry. Throughout the 1980s, the new generation of powerful UNIX workstations and emerging 3D rendering was inevitably shifting the CAD software market to 3D and solid modelling. The first version of IGES (Initial Graphics Exchange Specification) had been published in 1980 but already the emerging shift to 3D CAD software using solid models (and also the need for such CAD software to manage product data such as material properties, surface finish, engineering tolerances, etc.) was creating a need for a new data exchange standard. In 1984, the PDES (Product Data Exchange Specification) initiative was started in Europe to address the new needs.

CAD SYSTEMS

CAD is used in many ways depending on the profession of the user and the type of software application adopted. There are several different types of CAD. Each of these different types of CAD systems requires the operator to think differently about how they will use them and how they must design their virtual components in a different way for each. When CAD was first introduced for the mainstream AEC/FM industry in the 1980s, two types of systems emerged: entity-based CAD (such as AutoCAD) and object-based CAD (such as ArchiCad). The differences between the two systems are responsible for the fundamental difference between their applications.

ENTITY-BASED CAD

Entity-based or entity-orientated CAD uses vector graphics such as dots, lines and arcs, etc., to form the design – a series of lines, for instance, would form the design of a wall. This will be discussed in Chapters 3 and 4.

There are many producers of 2D systems, such as AutoCAD, which are based on the entity-orientated approach, including a number of free/open programs. These provide an approach to the drawing process without all the fuss over scale and placement on the drawing sheet that accompanies hand drafting, since these can be adjusted as required during the creation of the final draft.

3D wireframe is basically an extension of 2D drafting. Each line has to be manually inserted into the drawing. The final product has no mass properties associated with it and cannot have features such as holes directly added to it. The operator approaches these in a similar fashion to the 2D systems, although many 3D systems allow the use of the wireframe model to make the final engineering drawing views.

3D 'dumb' solids are created in a way analogous to manipulations of real-world objects. Basic three-dimensional geometric forms (prisms, cylinders, spheres, etc.) have solid volumes added or subtracted from them, as if assembling or cutting real-world objects. Two-dimensional projected views can easily be generated from the models. Basic 3D solids don't usually include tools to easily allow motion of components, set limits to their motion, or identify interference between components.

OBJECT-BASED CAD

Conversely, object-based or object-orientated CAD uses complete parametric objects like walls, windows, etc., to form the design. This is covered in Chapters 5, 6 and 7.

This form of modelling lends itself easily to actual construction, and subsequently to CAM tools. 3D parametric solid modelling requires the operator to use what is referred to as 'design intent'. The objects and features created are adjustable. Any future modifications will be simple, difficult or nearly impossible, depending on how the original part was created. One must think of this as being a 'perfect world' representation of the component. If a feature was intended to be located from the centre of the part, the operator needs to locate it from the centre of the model, not, perhaps, from a more convenient edge or an arbitrary point, as he could when using 'dumb' solids. Parametric solids require the operator to consider the consequences of his actions carefully.

MODERN CAD APPLICATIONS

Entity-based CAD systems, and especially AutoCAD, saw the greatest uptake by industry, as the applications could be handled by the computer processors of the 1980s. However, the way in which the actors in the building process communicated didn't change; it was still done largely with 2D drawings. The only difference now was that the lines were drawn using a computer instead of pencil and paper. CAD's greatest benefit was its ability to copy, rotate and scale parts of the design without having to draw again from the beginning. Its introduction virtually eliminated the use of pen and paper in architectural firms that were still using traditional methods. Its ease of uptake led to a trend of downsizing in drafting departments in many small to medium sized AEC/FM organisations. As a general rule, one CAD operator could readily replace at least three to five drafters using traditional pen and paper methods. This trend mirrored that of the elimination of many office jobs traditionally performed by a secretary as word processors, spreadsheets, databases, etc., became standard.

Although the commercial development of the two types of CAD systems (entity-based and object-based) began at approximately the same time in the mid-1980s, it wasn't until the late 1990s and the beginning of the twenty-first century that computers, and especially their graphic cards, improved significantly. Therefore, it wasn't until this period that it became possible to visualise object-oriented CAD in an acceptable way. In 2002, Autodesk bought Revit Technology Corporation in order to develop a suite of programs based upon parametric object-oriented CAD. During the same period the concept of using a Building Information Model (BIM) was also introduced. A building information model is not only rich with geometric 3D data about the building; it also has additional information that can be useful throughout the building's whole lifecycle. This additional information in computer models has made it possible to talk about 4D (3D + time/scheduling; see Chapter 5), and even nD (involving an infinite number of information properties; see Chapter 7). An additional benefit of using BIM is the possibility of viewing the model in real-time. Only the future will reveal the extent of the 'n' in nD, and this will become clearer when the building industry begins to discover and exploit the full extent of BIM potential.

Discussion

- Does your experience of CAD aid the design process or the drafting process?
- What are the benefits of CAD?
- What are the drawbacks of CAD?

1.4 CAD technology

HARDWARE AND OPERATING SYSTEM (OS) TECHNOLOGIES

Today, most CAD applications are designed for Microsoft (MS) Windows-based PCs. Some CAD systems also run on one of the Unix operating systems and with Linux. Mac OSX is also a viable platform for CAD applications, with Graphisoft's ArchiCAD, currently the most popular CAD system for this operating system, and, more recently, AutoCAD for Mac, which has been reintroduced by Autodesk. Some CAD applications such as QCad, NX or CATIA V5 provide multiplatform support, including MS Windows, Linux, UNIX and Mac OSX. Generally, no additional hardware is required, with the possible exception of a good graphics card depending on the CAD software used. However, for complex product design, machines with high speed (and possibly multiple) CPUs (computer processing units) and large amounts of RAM are recommended.

COST

The cost of hardware and software has had a profound effect on the widespread application of CAD. As CAD packages started to emerge and develop in the mainstream market in the 1980s and 1990s, latest advances were quite expensive, and small and even medium sized firms often could not compete against larger firms who were able to use their computational edge for competitive purposes. Nowadays, both hardware and software costs have come down. Even high-end packages work on less expensive platforms and some even support multiple platforms. Many software vendors offer free student licences and plug-ins. The costs associated with CAD implementation are now more heavily weighted towards the costs of training in the use of these high-level tools, the cost of integrating CAD/CAM across multi-CAD and multi-platform environments and the costs of modifying design work flows to exploit the full advantage of CAD tools.

INTERFACE

The human-machine interface of CAD applications is generally via a computer mouse, but can also be via a pen and digitising graphics tablet. Manipulation of the view of the model on the screen is also sometimes achieved by use of a spacemouse/spaceball. Some systems also support stereoscopic glasses for viewing the 3D model.

Discussion

- What is the purpose of CAD?
- What is the difference between entity-based and object-based CAD?

Chapter 2

Project and product modelling

2.1 Introduction

This chapter provides a succinct overview of the AEC/FM industry in order to establish an understanding of the development and importance of CAD. It provides a backdrop for the role of CAD in the design and construction process, and serves as a background for the later chapters in this book.

The construction industry worldwide has been continuously criticised for its less than optimal performance in several government and institutional reports: in the UK these have included Emmerson (1962), Banwell (1964), Gyles (1992), Latham (1994) and Egan (1998), to name but a few. As modelling is promoted as a way forward, this chapter looks at the various techniques that can be used for both process and product modelling.

Process modelling and product modelling are terms that are often associated with CAD. They are essentially differing perspectives of co-ordinating construction information, which is necessary for both comprehensive and coordinated CAD. This chapter will examine both techniques. It will begin by defining what a process is, before describing the various approaches to modelling processes and illustrating the development of generations of process maps. It will also provide examples of processes used in the management of construction. The second part of the chapter will cover product-modelling paradigms.

2.2 Why a process model?

In order to improve something it is necessary to know in advance what its current state is. Without knowledge of how the process looks and works today, it will be very difficult to know what improvement initiatives can be applied and the extent to which they will work.

In terms of CAD, process modelling the project has clear advantages. For instance, it can clearly highlight the communication and input relationships for all those who are involved in the design process.

ADVANTAGES OF PROCESS MODELLING

Process modelling involves producing a picture or map or a model that helps to make work visible. Increased visibility improves communication and understanding, and provides a common frame of reference for those involved with the work process; it should be the first step in any improvement activity. Process modelling is also a tool that provides a means of communicating complex functions in a form more easily understandable by people, thus enabling individuals to work together more efficiently. Modelling enables the formalisation of processes which in turn allows people to operate in a standardised manner. Maps are often used to show how work currently gets done, representing a snapshot in time that shows the specific combination of the functions, steps, inputs, and outputs. Process models are also an aid to the understanding of the way in which processes dynamically work with information.

Analysis of the processes which the maps represent can help to increase client/customer satisfaction in the project by identifying actions to decrease defects, reduce costs, establish customer-driven process-performance measures, reduce non-value-added steps, and increase productivity. Organisations can also use maps and flow charts to show how they want work to be done. By examining a map of current process performance in the light of client/customer requirements and data on sources of client/ customer-perceived value, organisations can draw a different picture to help them in illustrating the pathways that will be created to provide value to their customers. In addition to using process maps to show how work currently gets done or how they want work to be done, organisations can also use process maps to: orient new employees; evaluate or establish alternative ways of organising people to get the work done; quickly get up to speed on what individual groups, teams, or departments provide to the rest of the organisation and vice versa; identify improvement opportunities; and evaluate, establish, or strengthen performance measures.

2.2.1 The traditional design and construction process model/ RIBA Plan of Work

Process modelling has often been cited as a learning point for other industries to improve their practice, in particular the construction industry (Howell, 1999). The UK construction industry is under increasing pressure to improve its practices (Hill, 1992; Howell, 1999). It has been continuously criticised for its less than optimal performance by several government and institutional reports, such as Emmerson (1962), Banwell (1964), Gyles (1992), Latham (1994) and, more recently, Egan (1998). Most of these reports conclude, time and time again, that the fragmented nature of the industry, lack of co-ordination and communication between parties, the informal and unstructured learning process, adversarial contractual relationships, and lack of customer focus is what inhibits the industry's performance.

In 1959, the United Nations defined the building (project) process as 'the design, organisation and execution of building project' that has come to be recognised as 'normal practice in any country or region . . . it is characterised by the fact that all operations follow a set pattern known to all participants in the building operation' (United Nations, 1959). However, this description is essentially untrue today. The nature of the design and construction process has grown in complexity since the 1950s, thus leading to an increased number of actors in the project.

The term largely associated with the 'traditional building process' today usually refers to the practice where, upon perceiving a need for a new facility, a building client approaches an architect/engineer to initiate a process to design, procure and construct a building to meet his/her specific needs. The process, in turn, almost invariably consists of the project being designed and built by two separate groups of disciplines who collectively form a temporary multi-organisation for the duration of the project: the design group and the construction group (Mohsini and Davidson, 1992). The design group, typically, is coordinated by an architect/ engineer. Depending upon the circumstances of the project at hand, it may also include other design professionals and specialists such as engineers, quantity surveyors, etc. The principal function of this group is to prepare the design specifications of the work and other technical and contractual documents. The construction group, on the other hand, is usually coordinated by the main contractor and consists of a host of subcontractors and suppliers/ manufacturers of building materials, components, hardware and subsystems. This group is primarily responsible for the construction of the building project.

The two groups typically do not work coherently together (Kagioglou et al., 1998a, b and c). The design activities in construction are usually isolated from the realities of the real issues facing production, as each function is expected to play a specific and limited role in any phase, thus contributing to the industry's problems, as highlighted by the many governmental and industrial reports (Emmerson, 1962; Banwell, 1964; Gyles, 1992; Latham, 1994; Egan, 1998). This factor has contributed to the problems of construction with poor supply-chain co-ordination, fragmented project teams and adversarial relationships (Mohsini and Davidson, 1992).

Traditionally, a whole host of construction specialists, from both groups, are involved in instigating the design of modern buildings. With so much information, and from so many experts, it becomes very difficult to coordinate the information and to visualise any changes applied to the design, and this subsequently impacts on the time and cost of the project. Changing and adapting the design, planning schedules and cost estimates to aid decision-making can be laborious, time consuming and costly. Furthermore, each of the design parameters that the stakeholders seek to consider will have a host of social, economic and legislative constraints that may be in conflict with one another. As each of these factors vary – in the amount and type of impacts they can have – they will have a direct impact on the time and cost of the construction project. The criteria for successful design therefore will include a measure of the extent to which all these factors can be co-ordinated and mutually satisfied to meet the expectations of all the parties involved.

Specialist design criteria input is usually undertaken in a sequential step-by-step fashion, whereby the design undergoes a number of changes; after satisfying the legal requirements, it then proceeds to the next consultant who in turn makes a number of design recommendation changes. Design changes are made in isolation from each other in an over-the-wall manner, whereby each discrete change pays little or no regard to the next (see Figure 2.1). Therefore, it is often difficult to balance the design between aesthetics, ecology and economy – a three-dimensional view of design that acknowledges its social, environmental and economic roles – in order to satisfy the needs of all the stakeholders. These problems are mainly attributed to the vast amount of information and knowledge that is required to bring about good design and construction co-ordination and communication within a traditionally fragmented supply chain. The complexity of the problem increases with the fact that this information is produced by a number of construction professionals from different backgrounds. Therefore, without effective implementation of processes to control and manage this information, the problem will only intensify as construction projects become more and more complex, and as stakeholders increasingly enquire about the performance of buildings (sustainability, accessibility, acoustic properties, energy efficiency, maintainability, crime deterrence, etc.). Hence the need for improved project modelling.

The success of construction projects largely hinges on both the efficacy of the project process and the team enacting the work (Sidwell, 1990). The construction team is a living organism, usually formed from various organisations for the temporary duration of the project (Lingle and Schiemann, 1996). Although they have different priorities and capabilities, they are expected to work cohesively together from the outset. However, cultural issues between project team members have often been cited in the literature as limiting the project's success (Sidwell, 1990), and team-building exercises are commonly introduced to counterbalance this. In addition, the number and subsequent variations of design and construction processes creates a much deeper problem for the efficacy of the project (Howell, 1999), and there is no standard project process in construction (Latham, 1994). This gives a clear indication of the problem facing construction: how are professionals meant to instinctively organise themselves into a team working environment when the process is full of infinite variation and their roles and responsibilities vary from project to project. What part does CAD play in this process?

RIBA PLAN OF WORK

The Royal Institute of British Architects' (RIBA) Plan of Work (RIBA, 1997) was designed to represent the UK design and construction project process. The model (see

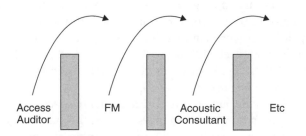

Figure 2.1 Sequential over the 'brick wall' approach to design and construction

Figure 2.1) was originally published in 1963 as a standard method of operation for the construction of buildings, and it has become widely accepted as the operational model throughout the building industry (Kagioglou et al., 1998a, b and c). The Plan (see Figure 2.2) represents a logical sequence of events that should ensure that sound and timely decisions are made during the course of the project. It suggests that all the decisions, set out or implied, have to be taken or reviewed (RIBA, 1997), and it is anticipated that the model will only need slight adjustments depending upon the size and complexity of the project. The project progresses from inception to feedback, that is from stages A to M, in a linear fashion requiring the completion of one stage before proceeding to the next. However, the design and construction process is essentially non-linear and cannot be viewed in such a functional fashion. This sequential flow only emphasises the divisions within the organisational structure of the industry and contributes to the problems of fragmentation, co-ordination and communication between project team members (Sheath et al., 1996), as highlighted earlier by many governmental and institutional reports (Phillips, 1950; Emmerson, 1962; Banwell, 1964; Gyles, 1992; Latham, 1994; Egan, 1998). The Plan was designed from an architectural perspective, which has in some way restricted its application to specific forms of UK construction contracts and it is increasingly inappropriate for the newer types of contracts being used both in the UK and elsewhere, such as 'partnering' frameworks. Given the complexity of clients' needs, the advancement in technology and materials, and the increased number of design and construction professionals, the need for improved process modelling in the design and construction industry is paramount.

In terms of CAD, it is only used in a rudimentary way at Stage C, Feasibility of the Plan of Works, and it largely comes into its own at Stages D and E. Prior to Stage C, hand-drawn sketches are still commonplace within the industry; and from Stage F onwards, CAD's role is rather limited. Once the CAD drawings have been prepared,

Pre-design	A B
Design	C D E
Preparing to build	F G H
Construction	J K L
Post-construction	M

Stage A: Inception Stage B: Feasibility
Stage C: Outline proposals StageD: Scheme design
Stage E: Detail design Stage F: Production info
Stage G: Bills of quantities Stage H: Tender action
Stage J: Project planning Stage K: Operations on site
Stage L: Completion Stage M: Feedback

Figure 2.2 RIBA Plan of Work

due to incompatibility issues between differing software packages, the design is often 'copied' by other design and construction professionals to outline their schemes for production (for example, building services engineering drawings). In short, there is often little or no interaction between differing CAD packages, and the essence of CAD can traditionally be seen as static.

This chapter continues by describing how CAD can be used to enhance the latter stages of a design and construction project by firstly defining its role in the design and construction process.

Discussion

- What are the drawbacks of the RIBA Plan of Work?
- How can the model be improved?
- What role does CAD play in an AEC/ FM project that you are familiar with? How could CAD be used to improve this project?

2.3 What is a process?

DEFINITION OF A PROCESS

Simply stated, the term 'process' has an input and an output, with the process receiving and subsequently transforming the input into the desired output (see Figure 2.3). A process can be visible, and at the same time, it can be invisible.

We all tend to do familiar tasks things in the same way, in a manner we are used to, and do not reflect upon the fact that 'now I am performing an activity' or 'now I have completed this task'. However, in order to model a task or a process, we need to describe the 'what happens'. Often, nouns, verbs and adjectives are used to depict a process (Lundgren, 2002). The noun usually refers to a person, place or object; a verb being a word or a phase that describes a course of events, conditions or experiences; and the adjective specifying an attribute of the noun (see Figure 2.4). There is a flow

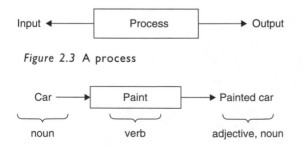

Figure 2.3 A process

Figure 2.4 Description of a process

relation between the noun, the verb and the adjective – a car is painted and the result is a painted car.

This is a mere simple descriptor of a simple process, and in fact a process in a project context is generally complex. Every successful project needs a 'formal blueprint, roadmap, template or thought process' (Cooper, 1994). Table 2.1 illustrates the various approaches to defining a 'process.'

2.4 Approaches to process modelling

A process model is a representation of a set of components of a system or subject area. It is developed for understanding, analysis, improvement or replacement of the system. Systems are composed of interfacing or interdependent parts that work together to perform a useful function. System parts can be any combination of things, including people, information, software, processes, equipment, products, or raw materials. The model describes what a system does, what controls it, what things it works on, what means it uses to perform its functions, and what it produces.

An understanding of processes can be reached in different ways. The project process is often depicted/modelled to enhance team coordination and communication through simple mechanisms such as flow (see Figure 2.5) and Gantt charts (a flow chart that encompasses time; see Figure 2.6). There are many different communities involved in process modelling, such as managers and software developers, resulting in the existence of many different methodologies in use today. Each community seeks to

Table 2.1 Definition of a process

Author	Definition
Davenport (1993)	'A process is simply a structured, measured set of activities designed to produce a specified output for a particular customer or market' and it is the structure 'which an organisation follows that is necessary to produce value for its customers'.
Cooper (1994)	'Provides the thinking and action framework for transforming an idea into a product, and it can either be tangible or intangible, functionally based or organisationally based.'
Oakland (1995)	'The transformation of a set of inputs, which can include actions, methods and operations, into outputs that satisfy customer needs and expectations, in the form of products, information, services or – generally – results.'
Zairi (1997)	'A process is an approach for converting inputs into outputs. It is the way in which all the resources of an organisation are used in a reliable, repeatable and consistent way to achieve its goals.'
Bulletpoint (1996)	Suggests that regardless of the definition of the term 'process' there are certain characteristics that a process should possess: – Predictable and definable inputs – A linear, logical sequence of flows – A set of clearly definable tasks or activities – A predictable and desired outcome or result.

prescribe a set of graphical symbols or notation with defined syntax and semantics for its proper use. However, they all share the same central concepts such as the use of a rectangular box (sometimes with rounded corners) to represent functions and the use of arrows to represent the flow of inputs into functions and outputs out of functions. In order to model more complex scenarios of real-world phenomena, techniques such as IDEF0 (Integrated DEFinition language) and analytical reductionism/process decomposition are commonly used (Koskela, 1992). These are briefly introduced.

2.4.1 Flow charts

A flow chart is a graphic representation of the sequence of steps that make up a process. The use of flow charts is really a reinforcement of the fact that it easier to understand something presented graphically rather than when it is described. Put simply: 'A picture is worth a thousand words.'

There are many ways of drawing flow charts. The most basic way is simply to use different symbols to represent activities and arrows to illustrate the connections between activities. When it comes to the symbols used, there are a number of variants, ranging from complex shapes to simple boxes and lines. It is not worth claiming that one way is better than another. The important point is simply that the users must share a common understanding of the symbols. Some commonly used symbols are illustrated in Figure 2.5.

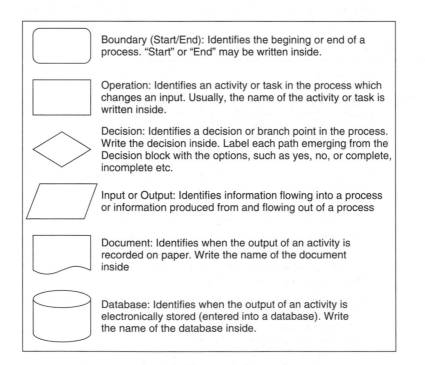

Figure 2.5 Common lexicon for flow charts

In addition, it is possible to indicate on the side of the symbols in the flow chart what resource or equipment is being used and under which conditions the activity is being performed. An example of a flow chart is shown in Figure 2.6. This chart could have been made more detailed; it merely illustrates the principles for drawing a flow chart.

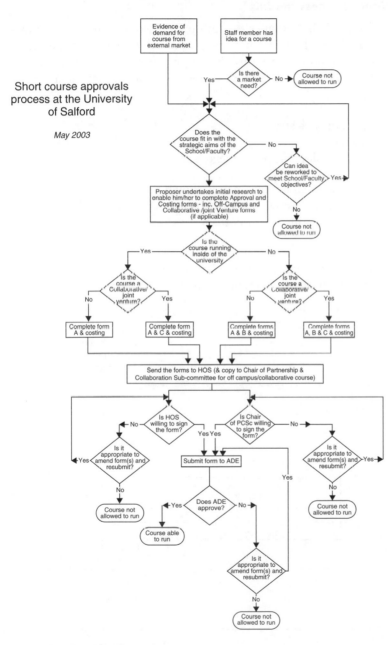

Figure 2.6 Example flow chart

It can, quite justifiably so, be argued that from the chart it is difficult to see who performs which tasks. This is possible in more sophisticated process model diagrams presented in the ensuing sections.

2.4.2 Cross-functional process models

A flow chart mainly describes the activities that are performed in a process. A cross-functional process map shows how business is conducted or work gets done in organisations/projects. They show the paths that inputs follow as they get transformed into outputs that add client's/customer's value. They show the steps that make up a process, as well as:

- the inputs and outputs of each step;
- the sequence of steps; and
- the people, functions, or roles that perform each step.

An example of a cross-functional process map is shown in Figure 2.7.

Cross-functional process maps show the value-producing chains of a business/project. They also depict the pathways to customer satisfaction by addressing the following:

- What steps are required to produce a particular output?
- What is the order in which the steps are performed?

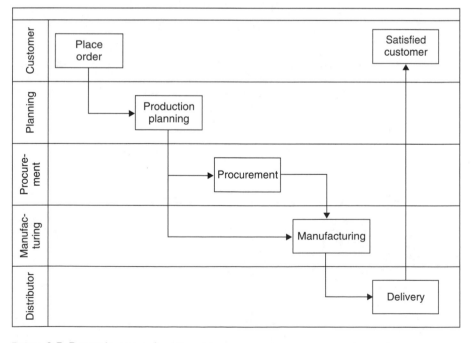

Figure 2.7 Example cross-functional process map

- Who (or which function) performs each step?
- What are the interfaces between functions?
- What are the inputs required and the outputs produced at each step of the process?

Cross-functional process maps often contain disconnects (missing or deficient inputs or outputs). Since cross-functional maps show what takes place inside one or more functions for a particular process, any disconnects that were present in the relationship map of those functions will also be present here. When a map is reviewed, inputs or outputs that do not feed into any other steps within the same function, nor into steps within other functions, may be discovered. Missing or implied steps, inputs, or outputs may also be discovered. Each of these is a form of disconnect that should be noted and resolved.

2.4.3 Analytical reductionism/process decomposition

Analytical reductionism/process decomposition involves breaking the process down into levels of granularity, as demonstrated in Figure 2.8, with the lower-level sub-processes further defining its corresponding upper-level process. It shares similarities with IDEF0 in that it breaks the parent process into subsequent, more detailed child (sub-)processes and then into procedures and activities. The level at which a process differentiates from a procedure is, however, still a topic of discussion in the process management field.

A process (Koskela, 1992; Cooper, 1994; Vonderembse and White, 1996):

- converts inputs into outputs;
- creates a change of state by taking the inputs (e.g. material, information, people) and passing it through a sequence of stages during which the inputs are transformed

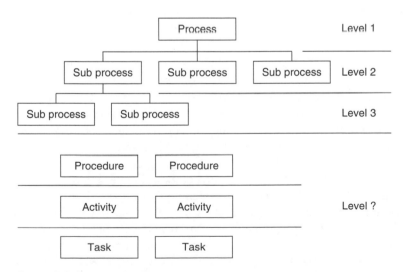

Figure 2.8 Process levels

or their status changed to emerge as an output with different characteristics. Hence processes act upon input and are dormant until the input is received. At each stage the transformation tasks may be procedural but may also be mechanical, chemical. etc.;

- clarifies the interfaces of fragmented management hierarchies;
- helps to increase visibility and understanding of the work to be done;
- defines the business/project activities across functional boundaries.

A procedure (Lee et al., 2001):

- is a sequence of steps. It includes the preparation, conduct and completion of a task. Each step can be a sequence of activities, and each activity a sequence of actions. The sequence of steps is critical to whether a statement or document is a procedure of something else;
- is required when the task we have to perform is complex or is routine and is required to be performed consistently;
- defines the rules that should be followed by an individual or group to carry out a specific task; their definition is usually rigid, leaving no opportunity for individual initiative;
- supports the process.

2.4.4 IDEF

During the 1970s, the US Air Force Program for Integrated Computer Aided Manu-facturing (ICAM) sought to increase manufacturing productivity through systematic application of computer technology. The ICAM program identified the need for better analysis and communication techniques for people involved in improving manu-facturing productivity and thus developed a series of techniques known as the IDEF family (IDEF, 2002):

- IDEF0, used to produce a 'function model' – a structured representation of the functions, activities or processes within the modelled system or subject area.
- IDEF1, used to produce an 'information model' – represents the structure and semantics of information within the modelled system or subject area.
- IDEF2, used to produce a 'dynamics model' – represents the time-varying behav-ioural characteristics of the modelled system or subject area.

In 1983, the US Air Force Integrated Information Support System program enhanced the IDEF1 information modelling technique to form IDEF1X (IDEF1 Extended), a semantic data modelling technique. Currently, IDEF0 and IDEF1X tech-niques are widely used in the government, industrial and commercial sectors, sup-porting modelling efforts for a wide range of enterprises and application domains. For the purpose of this chapter, IDEF0 will be described as it most closely relates to the 'functional' nature of the design and construction process.

The Integrated DEFinition language 0 for function modelling is an engineering technique for performing and managing needs analysis, benefits analysis, requirements definition, functional analysis, systems design, maintenance, and baselines for continuous improvement (IDEF, 2002). A function model is a structured graphical representation of the functions, activities or processes within the modelled system or subject area. IDEF0 models provide a 'blueprint' of functions and their interfaces that must be captured and understood in order to make systems engineering decisions that are logical, integratable and achievable, and provide an approach to:

- performing systems analysis and design at all levels, for systems composed of people, machines, materials, computers and information of all varieties – the entire enterprise, a system or a subject area;
- producing reference documentation concurrent with development to serve as a basis for integrating new systems or improving existing systems;
- communicating among analysts, designers, users and managers;
- allowing team consensus to be achieved by shared understanding;
- managing large and complex projects using qualitative measures of progress;
- providing a reference architecture for enterprise analysis, information engineering and resource management.

The result of applying IDEF0 to a system is a model that consists of a hierarchical series of diagrams, text and glossary cross-referenced to each other. The two primary modelling components are functions (represented on a diagram by boxes) and the data and objects that inter-relate those functions (represented by arrows; see Figure 2.9).

The boxes of the IDEF0 technique represent functions, defined as activities, processes or transformations. Each box should consist of a name and number inside the box boundaries; the name is of an active verb or verb phrase that describes the function; and the number inside the lower right corner is to identify the subject box in the associated supporting text.

The arrows in the diagram represent data or objects related to the functions, and do not represent a flow or sequence as in the traditional process flow chart model. They convey data or objects related to functions to be performed. The functions receiving data or objects are constrained by the data or objects made available. Each side of the function box has a standard meaning in terms of box/arrow relationships. The side of

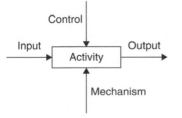

Figure 2.9 The basic concept of the IDEF0 syntax

the box with which an arrow interfaces reflects the arrow's role. Arrows entering the left side of the box are inputs; inputs are transformed or consumed by the function to produce outputs. Arrows entering the box at the top are controls; controls specify the conditions required for the function to produce correct outputs. Arrows leaving the box on the right side are outputs; the outputs are the data or objects produced by the function. Arrows connected to the bottom side of the box represent mechanisms; upward pointing arrows identify some of the means that support the execution of the function.

The functions in an IDEF0 diagram can be broken down or decomposed into more detailed diagrams, until the subject is described at a level necessary to support the goals of a particular project (see Figure 2.10). The top-level diagram in the model provides the most general or abstract description of the subject represented by the model. This diagram is followed by a series of child diagrams providing more detail about the subject. Each sub-function is modelled; on a given diagram, some of the functions, none of the functions or all of the functions may be decomposed individually by a box, with parent boxes detailed by child diagrams at the next lower level. All child diagrams must be within the scope of the top-level context diagram/parent box. In turn, each of these sub-functions may be decomposed, each creating another, lower-level child diagram.

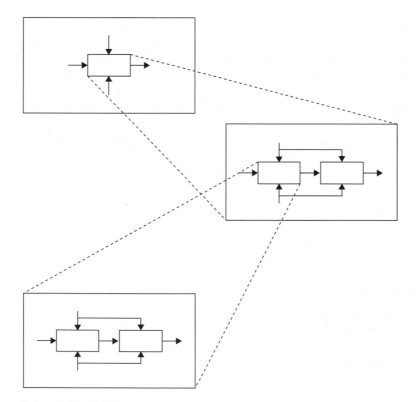

Figure 2.10 IDEF decomposition structure

2.4.5 Process modelling summary

Knowledge of how a process looks and works will help in identifying how to co-ordinate CAD and identify those areas in which to focus process improvement initiatives, thus providing the basis to then identify the extent to which these initaiatives are working once improved. Process modelling involves producing a picture or map or a model of organisational processes and should be the first step in any improvement activity. Thus, process modelling is an important part of project design and project redesign. It is a tool that provides a means of communicating complex functions in a form more easily understandable by people. It is not always possible to select all processes in a project for improvement and the key processes that have a significant impact on customer satisfaction should be selected. Popular techniques for process modelling include organisational or relationship diagrams, flow charts, cross-functional process maps, process decomposition and IDEF0 models.

Discussion

- What are the benefits of process modelling for CAD?
- What are the similarities and differences in the various process modelling techniques? Which are more suitable for improving the CAD process?

2.5 Product modelling

As discussed in the previous section, process modelling involves the modelling of processes in a project, and can often include the data and material that flows between them. Conversely, product modelling is used to model the elements specific to a product (CAD model) and the related process relationships; visual models are commonly produced through the mapping of conceptual data and process models and describe the information infrastructure of the product under development. The rapid prototyping of buildings/products using 3D/VR (virtual reality) technologies enable developers and clients to quickly assess and evaluate their requirements before committing fully to the project. This section of the chapter will consider aspects of the way information is used in product modelling and, by example, show specifically how IT is used to model the construction product.

Product modelling will be further discussed in later chapters, namely from Chapter 5 onwards.

Currently, the UK construction industry has not fully adopted and envisaged the benefits of using product modelling, unlike other industries such as the aerospace and automotive sectors. This is largely attributed to its deployment on an ad-hoc basis without context or framework, leading to the development of unreliable information models that become unusable over time. Thus, efforts and resources of product modelling are wasted. In addition, as discussed earlier, the construction industry is divided for historical rather than logical reasons. These divisions tend to reflect the traditional roles performed by the various disciplines and not the information required to complete the project. This leads to problems associated with information and project team integration. What is the solution?

It has recently been cited by leading researchers (Lee et al., 2003; Dawood et al., 2003; Fischer, 2000; Graphisoft, 2003; Rischmoller et al., 2000) that object technology, coupled with client server applications and the web environment, will provide the best way forward to enable project collaboration and information sharing by evoking a central project-based information database (building information/product model) and exchange between professionals. Up until the early 1990s Graphical schema languages such as Entity Relationship Diagrams, NIAM, IDEF1X were commonly used to undertake information modelling within the construction industry (Bjork and Wix, 1991; Rasdorf and Abudayyeh, 1992). Now, UML (Unified Modelling Language) has become more popular because of its wide use in the software industry. However, the use of such modelling techniques is not advocated as being appropriate for the industry, as they imply a separation between the data and the processes performed on the data. In order to overcome this problem, object-oriented models can be developed to describe the static information as well as the behaviour of objects. This has proven to be more advantageous as the resulting information model is richer and more natural, thus more usable for construction and other industries. This, it is anticipated, will enable effective co-ordination and communication of information amongst all project team members.

Unlike traditional data modelling techniques, the product-oriented paradigm models can be viewed as a collection of objects 'talking to each other using message sending'. The behaviour of one object may result in changes in another object. For example, if object 'column' has been moved, it should send a message to object 'beam' (to which it is related) to tell it to re-size itself reflecting the 'object' change. This way of modelling is very powerful and is peculiar to the object-oriented world. In such a world, objects can be composed of other objects. These objects can be images, speech, music or possibly a video. The object-oriented paradigm also supports the notions of encapsulation, abstraction, inheritance and polymorphism (Martin and Odell, 1992) that were considered as critical in handling the complex task of information modelling. Encapsulation permits objects to have properties (data) and actions (operation). For instance, an object 'beam' can have properties such as 'length,' 'width', etc., and behaviours or actions such as 'move beam', 'calculate load on beam', etc. Abstraction allows the analyst to abstract information according to requirements. For instance, the information about a beam can be abstracted in terms of properties, shape, materials, etc. Inheritance allows information in the parent object (beam) to be inherited by the child object (cantilever beam). Polymorphism allows objects to have one operation that can have different implementations. For instance, an operation 'calculate area' can be attached to an object called 'beam'. However, the implementations of this operation differ according to whether the beam is a 'rectangular beam' or 'T beam' etc.

Another major benefit of product orientation is the support of the notion of reusability. Using such a notion, integrated databases can be developed from re-usable object-oriented components that can be assembled as required. This is very similar to the way a building designer uses re-usable plans that can be configured to his/her requirements. The object-oriented paradigm also supports the notion of 'perspectives'. This notion allows the construction professional to view the information from his own perspective. For instance, the architect is interested in features such as colour, aesthetics, texture, etc., whereas the construction planner is interested in features such as time, resources, etc. To illustrate this, take the concept of a wall. This can be viewed from different perspectives. An architectural wall has attributes such as dimensions, colour and texture. A construction planner wall has attributes such as dimensions, time, etc. It is therefore logical to store common information such as length, width, etc., in 'wall' that can be inherited by the architectural wall, etc., through inheritance.

Product modelling is aimed at the identification of concepts/data – relationships between the concepts, attributes and operations that are to be supported by the database. This task should be done independently of any implementation platform. Figure 2.11 shows an illustration of an object model incorporating objects, relationships, attributes and operations

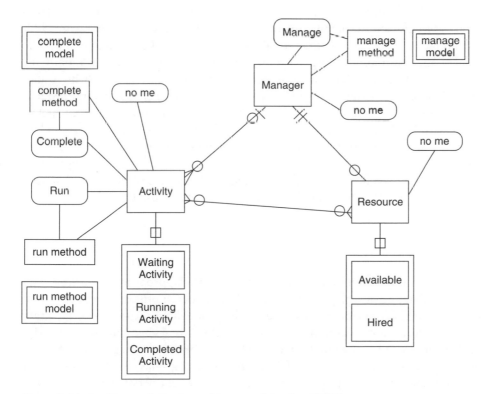

Figure 2:11 An illustration of an object model using IDEF0

2.5.1 Product data technologies

As described earlier in this chapter, the activities performed within a construction project can be modelled using techniques such as activity hierarchy, data flows, IDEF0 techniques and flow charts. These techniques describe the information flows between processes. This is useful in understanding how information is communicated between processes.

In the context of this chapter, Product Data Technology (PDT) refers to techniques of data modelling, data exchange and data management, which are aimed at the integration of product information through standard data models. Historically, the initial requirement for a standardised data model came from the need for different versions of CAD applications to share their graphic files. IGES (the Initial Graphics Exchange Specification) was developed as a protocol for this purpose. However, graphical and geometrical data is only part of the information required in a building project. IGES is not able to support the exchange of other types of data such as construction, thermal, light, etc. Therefore a new project, PDES (Product Data Exchange Specification), was proposed in the USA in the early 1980s to overcome these limitations. In the same period, similar efforts were made in other countries, for example the SET (Standard d'Echange et de Transfert) in France and the VDAFS (Verband der Deutschen Autombilindustrie Flaechen Scnittstelle) in Germany. In 1983, all these initiatives were coordinated into a major international programme under the umbrella of the International Organisation for Standardisation – the Standard for Exchange of Product data (STEP). Consequently, this became a comprehensive ISO standard (ISO 10303) for describing how to represent and exchange digital product information. In the construction industry, IFC (Industry Foundation Classes), which is compliant with STEP, was developed as a standard for exchanging building product data. IFCs are an interoperable data standard that is linked to any proprietary software application.

This will be discussed in more detail from Chapter 5 onwards. Product modelling is the foundation for BIM (building information modelling) used by the AEC/ FM industry.

2.6 Summary

This chapter has only been able to touch on the main approaches to process and product modelling, however, it does illustrate how critical both are to effective and efficient futures. The work on process and product modelling (primarily to integrate construction information) was initiated more than thirty years ago. However, fruitful practical results are only now slowly emerging because of the complex structure of the construction industry and because information technology has not been exploited properly. The inclusion of object types viewed from different perspectives but shared across different domains and abstraction levels is seen as a major step forward in integrating information throughout the construction industry. Such structuring is considered essential for the development of accurate, understandable models. This will be detailed further in the final chapter on nD modelling, which integrates both process and product modelling. As Cooper et al. (2004) have noted, we might imagine:

A system that given an idea, illustrates alternatives, illustrates constraints, enables the understanding of both quantitative (time, cost, legislation) and qualitative (aesthetics, usability) dimensions. Enables all the stakeholders to participate and allows users to virtually experience the product concept. The system will determine the build specification, the manufacturing resources, the production processes, provides the drawings, the tool sets. A system where we can use our knowledge, in conjunction with the other stakeholders, to achieve the best solution, at the right cost in a faster time, and in a sustainable manner . . . This is the Holy Grail for process and product modelling. The need will not go away, indeed as companies, systems products and markets get even more complex we need the models to guide and help us make decisions. However organisational behaviour illustrates that we are not automatons, we will never work to a detailed and prescribed process and procedure, when situations demand innovation, creativity and constant change to enable us to compete. Yet we do need systems to help us work through a complex world for the benefit of its inhabitants. The challenge is to understand what systems are the most appropriate, how we can best introduce them into organ-isations, and the impact that they will have on our work behaviour and the future of the organisations who use them.

Chapter 3

2D CAD

3.1 An introduction to 2D CAD drafting

As discussed in Chapter 1, Section 1.2, parts or products can be represented in two dimensions, and, as noted in the sub-section 'Entity-based CAD' in Section 1.3, vector graphics such as dots, lines and arcs can be used to form a design. The process of executing the above can be defined as 2D drafting. The term 'drafting' is more commonly used to describe 2D CAD, as tools that are used to support this process typically mimic the manual process of drafting (drawing by hand) as opposed to aiding the actual process of design. Drafting is also defined as a way to descriptively deliver an idea with the use of drawings and/or illustrations that show the process of creating the idea and bringing such an idea into reality. To avoid confusion, the term '2D CAD drafting' will be used to refer to software-based 2D, and '2D manual drawing' will be used to refer to hand drafting.

This form of drafting has a historical past which was initially executed manually prior to the invention of CAD systems and applications. Each particular user can either choose to adopt the 2D drafting approach as an end representation in itself or as a step to proceed to 3D production.

However, for the adoption of 2D CAD drafting either as an end or as a step to 3D production, there are many advantages as well as disadvantages involved. It is important to point out that some factors could be a disadvantage to some users but could end up as an advantage to others as a result of peculiar context in which the 2D drafting is used.

3.2 The history of 2D CAD drafting

The process of drafting has not changed significantly over thousands of years except for the tools that are used to create a design and the means of preserving designs. Until the 1800s, animal skin was used to preserve a design because it did not shrink as paper did. Over the past 75 years, electronics have been increasingly used in the process of drafting and, more recently, the computer has been used to enhance the art of drafting

2D drafting has been used since the advent of the CAD tool. The first CAD systems served as mere replacements of drawing boards, the result of an attempt to digitise the hand-drawn 2D process. It was more than 2,300 years after Euclid that the first true CAD software, a very innovative system (although of course primitive compared to today's CAD software) called 'Sketchpad', was developed by Ivan Sutherland as part of his Ph.D. thesis at MIT in the early 1960s (see Table 1.1).

The CAD designer then worked in 2D to create a technical drawing consisting of 2D wireframe primitives (i.e. line, arc, etc.). The productivity of design departments increased, and as a result of the advantages provided by this method of drafting, it was widely adopted, with different software applications emerging to meet with its growing demand.

Discussion

- What are the advantages and disadvantages of 2D CAD drafting over 2D (hand) drawings?
- What is the difference between 2D and 3D?

3.2.1 The purpose of 2D drafting

Human imagination can be creative in a bid to solve a problem or to answer a question. This creativity usually involves a process which can span a long or short period of time. In order to document this process there is a need to recall information, but unfortunately the human memory is quite limited with regards to this, hence the need for documentation. Documentation of a system in a text format can involve huge volumes of paperwork, with limited information describing the creative product; on the contrary, drawings can provide more information within a limited amount of time. For the drawn idea to be available in a format suitable for production, dimensional and content accuracy is important, hence the need for 2D drawing and the use of CAD to increase this attribute exponentially.

There are numerous different work disciplines for drafters and designers where there is the need for quick and accurate representation of an intended product prior to production.

Everything that a person can hold in their hand, feel or touch has some type of drawing or illustration created for the production of that product. All of the world's greatest inventions, thoughts and improvements began with a drawing. These drawings are predominantly worked into the production process using CAD software.

Some of the key purposes of drafting can include:

i)   Preservation of illustrations.
ii) Assisting engineers and designers in a wide variety of industries to design and manufacture physical products ranging from buildings, bridges, roads, aircraft, ships and cars to digital cameras, mobile phones, TVs, clothes.
iii) Achieving less design time.

iv) Easier visualisation process.
v) Ensuring and maintaining high and consistent accuracy levels. Automated CAD drafting creates high-quality designs accurately and faster.

It is also important to note that in CAD drafting, the functionality of the software can be enhanced by integrating it with other useful applications.

Discussion

*   What 2D CAD drafting packages do you use? What reasons governed your choice over other packages?

3.2.2 The advantages of 2D CAD drafting

2D CAD drafting can be viewed as advantageous when executed with CAD software. Some of the most common 2D CAD drafting advantages are:

1. Product documentation: For a product to be actualised there needs to be a form of documentation which will serve as a guide to the production process.
2. Standardised multiple production: Where there is a need for reproduction, the 2D draft will serve as a guide to aid the production of more of the same item with the same standards.
3. Product development: 2D drafting also helps in the development of a product because the draft will tend to reveal design flaws which can be corrected and documented.
4. No communication or a language problem: Due to the nature of the final output in a drafting process the drawings produced can be understood to a large extent by industry professionals with different language backgrounds.
5. Interoperability: In the event of information/file exchange between allied industry professionals within a particular project, 2D CAD drafting will most likely enhance collaboration and information sharing among participating professionals. This, however, is dependent on the interoperability of the software adopted by these different practitioners. Interoperable software can be described as non-proprietary software. Some 2D software programs such as AutoCAD can be saved in file formats such as dxf which in turn can be viewed by a lot of other dxf format-compatible CAD software programs.
6. Increased efficiencies: During the process of CAD drafting, efficiencies can be achieved as a result of some of the functions present in most 2D packages. The 'copy', 'mirror', 'hatch' and 'array' tools which are present in most 2D packages and integral to the drafting process can potentially save cost and time. This saving could be in the order of days, weeks or even months when dealing with relatively large projects (see Figure 3.1 (a) and (b)). If it took X amount of time to draw the

cross in Figure 3.1 (a) manually, it is expected to take 16 times that amount of time to draw 16 of them (Figure 3.1 (b)) manually. However, using 2D CAD drafting and applying the tools mentioned above, it will take just the time it takes to copy in order to produce 15 additional crosses (Figure 3.1 (b)). This will result in increased efficiency.

7. Reduction in time and costs: With the CAD drafting capabilities illustrated in Point 2 above, the resulting efficiencies can be attributed among other factors to savings in cost and quicker turnaround times.

8. High quality drawings with accurate dimensions: As a result of the interface customisation capability and the zoom functions, it is easier to produce drawings of high quality. Each user of 2D CAD software is able to customise the tools in such a way as is suited to the individual user in order to achieve maximum production quality (see the example in Figure 3.2). Also the zoom function helps to magnify details in order to ensure the accuracy of smaller details both in design and dimensions (see example in Figure 3.3).

9. Competitive pricing: As time is a crucial factor in pricing a draft or design project, the time advantage of the CAD draft goes a long way in providing a pricing advantage and the enhanced accuracy assures a high-quality execution which also contributes to the pricing advantage. This helps clients to drastically reduce their operating expenses by paying for less draft turnaround time.

10. High data security: 2D drawings produced using CAD software can be secured by saving in non-editable formats such as a PDF. Files can also be saved in password-secured interfaces or in personal/protected drives.

11. Making revisions: CAD 2D drafting can be altered very quickly without having to erase unwanted drawings and redrawing the desired changes. Any mistake or

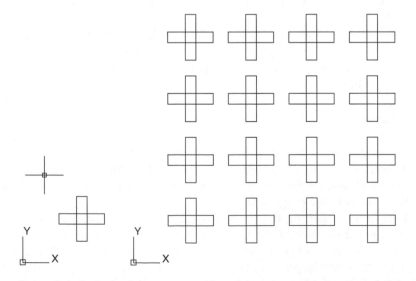

Figure 3.1 A single drawn cross object (a) and a multiplication of the same object (b) using either the copy, mirror, or array tools.

revisions can, for example, be adjusted quickly without the need to redraw the
entire drawing; with the push of a button during drafting, lines and shapes can
be easily and quickly altered. This process enhances neater and quicker 2D
drawings.

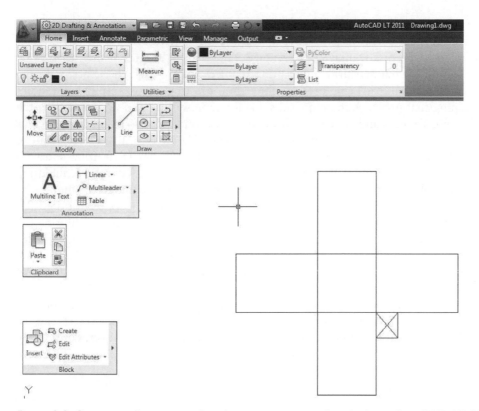

Figure 3.2 Customised user interface by repositioning of tools from AutoCAD LT 2011
ribbon.

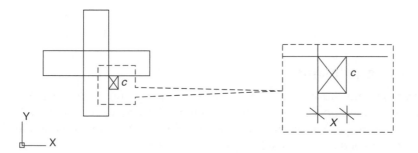

Figure 3.3 (a) showing a single drawn cross object with another object 'c' at the
lower right angle of the cross. 'c' was zoomed in (b), providing a larger
image for a more accurate visualisation and dimensioning of 'x'.

3.2.3 The disadvantages of 2D CAD drafting

There are fewer identifiable disadvantages of 2D CAD drafting over 2D manual drawing/drafting, and some of which are directly linked to the areas of advantages listed in Section 3.2.2. Key disadvantages are as follows:

1. 2D CAD drafting can be time consuming especially when tight production deadlines have been set. However, the advent of the CAD software has to a large extent eliminated this disadvantage (see Section 3.2.2, Point 6, 'Increased efficiencies').
2. Training: 2D manual drawing/drafting requires training which is specific to the sector in which it is being used. Architectural drafting, for example, will require training that provides the required skills for representing and understanding the drafting symbols and concepts. This will also be applicable to professions such as the engineering, surveying and creative design. 2D CAD drafting also requires additional training in skill acquisition on selected 2D CAD drafting software programs. Such training can be cost and time intensive and will require updating as new releases of those particular 2D software programs come to the market.
3. Loss of work as a result of not saving: on all too many occasions drafted 2D drawings have been lost by CAD users as a result of not saving them during the drafting process. It is important not to underestimate the significance of this oversight as hours of work can be lost at the last moment and lead to the contract being compromised or even terminated (dependent on the conditions outlined in the contract).
4. Interoperability: in terms of information exchange, the effective collaboration among allied industry professionals working on a particular project can be reduced if editable 2D files can't be exchanged effectively. However, considering the current positive trend in drawing file exchange technology; this disadvantage is poised to becoming an advantage.

Discussion

*   What advantages and disadvantages do you consider critical to 2D CAD drafting?

3.3 2D drafting principles

Although drafting conventions may vary slightly, the broad principles of 2D CAD drafting are always the same: suitability for the intended purpose, accuracy, legibility and neatness, economy in time and labour. All the above principles should be observed during geometry creation, editing and manipulation.

Subsequent sections will throw some light on the basic 2D drafting process and highlight some generic exercises for 2D CAD drafting which apply to the majority of industrial CAD software programs.

3.3.1 Creating a new drawing

In order to start a drawing project in a CAD environment, there are some important steps to follow. It is important to point out that some of these steps should be taken in a particular sequence while some can be applied at a later stage in the drawing process.

DRAWING FILE CREATION

A new drawing file can be created in most industrial CAD software programs, such as AutoCAD, ArchiCAD or Revit, by creating an entirely new sheet or using an already provided template. From the pull down menu, select 'save as' and assign a file name and save in prepared folder.

DRAWING UNITS

The drawing units, or drawing scale, can be defined by the user. The desired units can be in centimetres, millimetres, inches, etc. Industry standards with regards to drawing units vary across different professions and geographic locations. Look for 'drawing set up' in your CAD software either from the menu or tool bar. AutoCAD provides a command line at the bottom of the screen where you can type the command 'units', and press 'enter'. It is also important to set the desired precision of the drawing if your CAD software permits it.

DRAWING GRID AND SNAP

The majority of 2D CAD software programs support the grid system. The drawing grid is a regular pattern of dots displayed on the screen which acts as a visual aid; it is the equivalent of having a sheet of graph paper behind your drawing on a drawing board. You can control the grid spacing, so it can give you a general idea about the size of drawn objects. It can also be used to define the extent of your drawing. A grid can be displayed over the area defined by the limits. Some CAD tools provide a facility where objects are snapped to grid points or other objects within the drawing. Dimensions are assigned to grid spaces. To display the grid, or to activate snap, use the buttons along the bottom of the screen or the command line 'grid' command.

DRAWING LIMITS

'Drawing limits' is used to define the extent of the grid display and to toggle 'limits mode' which can be used to define the extent of your drawing. The grid is displayed within a rectangle defined by two pick points or co-ordinates. It is important to define the drawing space prior to commencement of drawing; this is a feature present in AutoCAD, which provides a command line at the bottom of the screen where you can type the command 'limits' and press 'enter'. Limits assignment will provide an idea of what scale to adopt for each drafting project embarked upon. However, the final scale to be printed is assigned at the time of printing.

3.3.2 Layering

Layering is a concept that is not limited to CAD programs but is also found in photo editing software. Layering can be referred to as a process of creating different work layers within a single drawing in order to help organise complex drawings. These layers are created according to each user's choice and a significant number of layers can be created at any one time.

CREATING A NEW LAYER

A new layer can be created by following the instructions provided for any specific software which allows layers. However, it is important to understand the basic principle of creating layers, which is that each layer should, if possible, be assigned a unique name, colour, linetype, line weight and a brief description if the software being used permits.

ASSIGNING AND HANDLING LAYERS

In CAD software programs which have layering functions, layer assignment and handling varies from one program to another, and it is advisable to read and follow the instructions provided for each program. Names based on elements within the drawing such as walls, windows, etc., are assigned to each of these layers for quick identification. Each of these layers can be switched on and off (in AutoCAD they can also be frozen and thawed).

It is important to know that for large projects, layering can greatly aid efficiency. For example, (i) if sanitary fittings in a large project have been assigned a layer, and it is necessary to delete them all, switching off all other layers will make it possible to display only the sanitary fittings and using the 'select all' and 'delete' commands will delete them within seconds; (ii) when engaging in a drawing which involves sending printouts to various allied professionals, switching off unwanted layers and allowing only that of the target profession to be printed also enhances organisation and hence efficiency; (iii) when working with a maze of lines and shapes in larger projects, layering enables the user to reduce the number of these lines and shapes by switching off other layers, allowing just the layer to be edited.

3.3.3 Drawing 2D lines and shapes

DRAWING LINES

Depending on the type of 2D CAD software used, lines can be drawn by using coordinates or by freehand, guided orthogonally by constraints provided by the software.

Coordinates can be relative or absolute. Using relative coordinates is when you click on your 'line' tool, click your 'start point' and apply coordinate values relative to this start point. The use of absolute coordinates, however, is when coordinate values are applied to both the start point of the line as well as the end point, based on the set limits of the drawing sheet.

When using the freehand approach in the majority of 2D industry-standard CAD software programs, the line command allows you to create straight lines, one after the other (chained). You specify the first point as one end of the first line, then the next point, which will be the end point of the first line and the start point of the second line, and so on. When multiple lines are linked at their ends and can be selected as a single line some CAD software such as AutoCAD refers to the line as polyline.

EDITING LINES

In order to use lines effectively for 2D drafting, the lines should be editable. Some of the editing functions involve:

a) Trimming: This is where lines are cut; especially unwanted parts of a particular line. Polyline can also be trimmed in ways similar to ordinary lines, the difference being that since the polyline is one continuous multiple line, all connected parts of the line on the side to be removed will be trimmed off when the command is applied.

b) Extending: Lines can be extended either by manually dragging the end or, in some software programs, selecting a target line along the path of the line to be extended and then clicking on the line after clicking the extension command.

c) Erasing: Lines can be erased by using the erase function of the CAD software being used. However, in the majority of CAD software programs, the 'delete' key on the keyboard can also perform the erase function as long as the objects to be erased have been selected.

d) Moving: Objects can be moved by using the move function of the 2D CAD software being used. However, some programs provide alternative ways of moving line objects such as typing some keyboard functions and selecting objects to move. Line objects can be moved using a freehand approach or by specifying distances, sometimes according to prescribed coordinates.

e) Copying: The copy command is an important function which is widely used during 2D CAD drafting. It can be applied in most CAD software programs in ways similar to the move command.

f) Mirroring: This function helps the users to mirror line objects or any group of line objects. Unlike copying, it generates its object by providing an inverted copy (symmetry) and can be used by architects and designers to replicate an inverted floor plan during a terrace design.

g) Arraying: Some 2D CAD software might not have the array function, but for those that do, the array tool copies an object in multiples in a pre-arranged format with a single command. Some arrangements are in a matrix format while some are radial.

h) Scaling: The scale function is present in the majority of 2D CAD software programs. This function helps its users to uniformly increase or reduce the size of a drawing while still maintaining the original dimensional relationships and geometries. This tool is applied either by using freehand or by assigning a scale factor. The assumption in most cases is that the original (pre-scaled) object is equivalent to '1' while the scale factor used to reduce the size of the object would then be a fraction of '1' and the scale factor used to increase it would be a multiple of '1'.

DRAWING SHAPES

Shape drawing is one of the basic capabilities of 2D software. However, the approach or ways in which to go about drawing these shapes can vary and is greatly dependent on the drawing/editing tools of the software being used. Some examples of common shapes are (a) squares/rectangles, (b) circles (c) triangles, or (d) polygons. Figure 3.4 (a) shows one way of drawing a square or rectangle by simply clicking on the rectangle tool and then clicking for the start point and clicking again for the end point. Coordinates can be used to assign the required dimensions; Figure 3.4 (b) shows four lines drawn and then the extensions trimmed off. The lines can also be snapped into position, which will eliminate the need to trim.

There are multiple ways of drawing the shapes listed above and this varies with individual users. Also, in the same way that lines can be edited, as shown in Section 3.2.3, shapes can also be edited.

3.3.4 Object properties

An object property is a function which describes the object in question within a CAD environment. In most cases the object properties can be edited except when the property in question has been locked by the user. In some CAD programs, there are alternative ways of altering some key properties of any given 2D object, such as its dimensions.

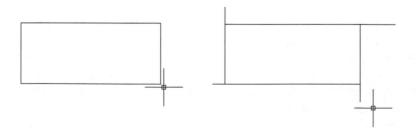

Figure 3.4 (a) A simple method of drawing a rectangle by using the rectangle tool; (b) an alternative way of drawing the rectangle by using ordinary lines and trimming off any excess.

DIMENSIONING AND ANNOTATION

Dimensioning in 2D CAD is straightforward if you follow a consistent procedure for setting things up at the start. The appearance or style of dimensions is highly configurable in most 2D CAD programs. The style that the user chooses can be selected from the software. Some CAD software packages utilise internationally recognised standards defined by organisations such as BSI (British Standards Institute), ISO (International Standards Institute) or ANSI (American National Standards Institute).

Some CAD software packages provide the user with the opportunity to simply check dimensions or actually to create dimensions for the drawing according to the selected style. Quite often, assigning dimensions and annotations are very similar.

3.3.5 Hatching

Hatching is a common function in most 2D CAD software programs. This feature is important as it helps the user give visual identity to a shape and provides options for different patterns.

CREATE HATCHING

Hatching can be created by clicking on the respective hatch function, selecting a closed geometry and then pressing 'enter'. However, some programs might have a different creation syntax, so it is important to read the tutorial for the specific software.

EDIT HATCHING

Occasionally the hatching created on a particular drawing might not appear to be what the CAD user expected, hence the need to alter it. Rather than deleting and going through the creation process again, the hatch editing tool present in some 2D CAD software can be used to edit the existing hatching by using some simple process as laid down in the user manual or tutorial.

3.4 2D CAD practical examples

The examples described below highlight the different functions involved in carrying out 2D drafting tasks. In these instances, the AutoCAD LT 2011 software program was used, however the examples focus on basic generic principles which can be applied to other 2D drafting software programs. The examples assume that preliminary drafting steps have been taken in the selected software for the functions mentioned in Sections 3.3.1 and 3.3.2, which involve creation of (i) the drawing file, (ii) the drawing units, (iii) the drawing limits, and (iv) the drawing grid. These can vary depending on which drafting software program is being used and the requirements specified in the software's tutorial.

Task

- What, and how many, different ways can you draw a square or circular shape in 2D CAD?

3.4.1 Example 1

In the first example, we are going demonstrate how to draw single rectilinear or straight lines and shapes such as squares and rectangles. Examples 'a' and 'b' below were carried out using AutoCAD LT 2011.

(A) CREATE LINES

To create a line in most 2D software programs, the following key steps are taken:

1. First step: Identify the line tool in the software being used.
2. Second Step: For most tools, you will need to click the line button or type in the line command and assign a thickness to it.
3. Third step: Click the start point of line '1' on the drawing or type in the co-ordinates of the desired start point.
4. Fourth step: Click the end point of line '2' on the drawing or type in the co-ordinates of the desired end point.

The above exercise can be repeated depending on the type of line being created. The lines created can be horizontal, vertical or inclined.

Edit lines: Lines created can be copied, moved, rotated, extended, trimmed, resized in thickness, or deleted. See Figure 3.5.

(B) CREATE RECTILINEAR SHAPES (SQUARES OR RECTANGLES)

To create a square or rectangle some key steps are taken:

1. First step: Identify the rectangle tool in the software being used.
2. Second step: For most tools, you will need to click the rectangle button or type in the rectangle command.
3. Third step: Click the start point of the rectangle '1' on the drawing or type in the coordinates of the desired start point.
4. Fourth step: Click the end point of the rectangle '2' on the drawing or type in the coordinates of the desired end point.

Drawing a square or a rectangle involves the same process, the difference being the dimensions assigned to the sides. The line tool can also be used to achieve this (see Example 1).

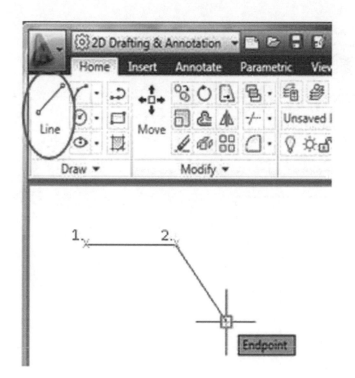

Figure 3.5 Drawing a line in AutoCAD LT 2011.

Edit lines: Rectilinear shapes created can be copied, moved, rotated, extended, trimmed, resized in thickness, or deleted. See Figure 3.6.

3.4.2 Example 2

In the second example, we are going draw single curved lines and shapes such as arcs, circles and ovals.

(A) CREATE ARCS

To create an arc in most 2D software programs some key steps usually have to be taken:

1. First step: Identify the arc tool in the software being used.
2. Second step: For most tools, you will need to click the arc button or type in the relevant command.
3. Third step: If you are drawing a 3-point arc, click the start point of the arc '1' on the drawing or type in the coordinates of the desired start point.
4. Fourth step: Click the mid-point of the arc '2' on the drawing or type in the coordinates of the desired mid-point.

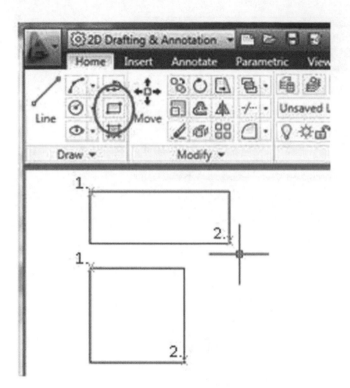

Figure 3.6 Drawing a square or rectangle in AutoCAD LT 2011.

5. Fourth step: Click the end point of the arc '3' on the drawing or type in the coordinates of the desired end point.

   See the relevant software tutorials for other ways of creating arcs.
   Edit arc: Arcs can be copied, moved, rotated, extended, trimmed, or deleted. See Figure 3.7.

(B) CREATE CIRCLES AND OVALS

To create a circle in most 2D software programs some key steps have to be taken:

1. First step: Identify the circle tool in the software being used.
2. Second step: For most tools, you will need to click the circle button or type in the relevant command.
3. Third step: Click on the drawing to position the centre of the circle '1'.
4. Fourth step: Assign the circle's radius or diameter or click the second point which locates the path of the circle '2'.

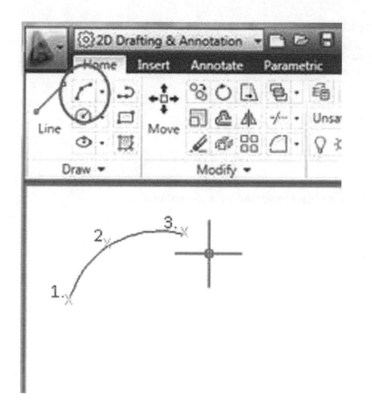

Figure 3.7 Drawing an arc in AutoCAD LT 2011.

For the oval shape, follow similar steps, but this time a third input is made which identifies the other radius of the shape. See the software tutorials for other ways of creating ovals.

Edit circles and ovals: Circles and ovals created can be copied, moved, rotated, trimmed, or deleted. See Figure 3.8.

3.4.3 Example 3

In the third example, we are going attempt hatching a square. However, the hatch area can take any shape as long as the boundaries form a closed loop.

(A) CREATE HATCHING

To create hatching in most 2D software programs some key steps have to be taken. This example is being carried out using AutoCAD LT 2011:

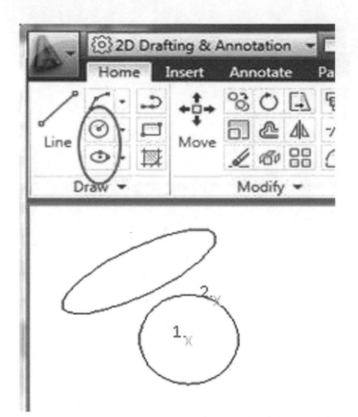

Figure 3.8 Drawing circles and ovals in AutoCAD LT 2011.

1. First step: You need to be familiar with the hatch tool in the software being used.
2. Second step: For most tools, you will need to click the hatch button or type in the hatch command.
3. Third step: For most drafting software, you will be provided with options to select what hatch type to adopt. Select the desired hatch pattern. You can rotate or scale this pattern by inserting numbers into the relevant boxes.
4. Fourth step: Select the area in the drawing to be hatched. For most drafting software programs, this area must be closed, which means that there shouldn't be a gap along its boundary. Then proceed to accept by clicking 'OK' or pressing the 'enter' button.

Edit hatching: To edit the hatching (for those software programs that permit it), click on the 'hatch edit tool', select the hatching to edit, assign new parameters and accept by clicking 'OK' or pressing the 'enter' button. See Figure 3.9.

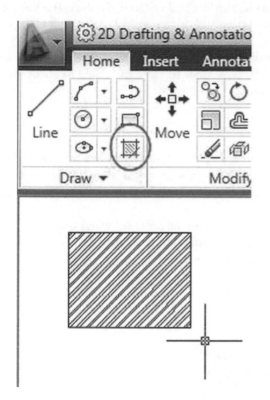

Figure 3.9 Hatching in AutoCAD LT 2011.

3.4.4 Example 4

In the fourth example, we are going draw an object which reflects a combination of rectilinear and curved lines and shapes with some hatch.

(A) USING SEPARATE EXERCISES IN EXAMPLES 1 TO 3 TO CREATE A HATCHED
WALL WITH A DOOR

For this exercise, the separate examples demonstrated in Sections 3.4.1 to 3.4.3 or Examples 1 to 3 are applied in the 2D drafting of a room corner and a door:

1. First step: Create the wall by either drawing horizontal and vertical lines or rectangles. The lines should be spaced appropriately to represent a wall.
2. Second step: A gap of appropriate dimensions should be provided for the door and this can be achieved by defining it from the start or later by trimming the line to allow for the door gap.

3. Third step: Draw a door symbol using the line and arc tool. To draw the line and arc, follow the steps in Examples 1(a) and 2(a); shown in Figure 3.5 and Figure 3.7 respectively.

4. Fourth step: Hatch the wall by following the steps shown in Example 3(a); see also Figure 3.9. To be able to hatch, remember to close the shape to ensure that the surrounding boundaries are closed.

3.5 Summary

In this chapter, we have determined that, historically, 2D drafting has been used since the advent of the CAD tool. The first CAD systems served as mere replacements of drawing boards and this was as a result of the attempt to digitise the hand-drawn 2D process.

2D drafting can be viewed as having numerous advantages when executed with CAD software. However, there are a few identifiable disadvantages with 2D CAD drafting when compared to 2D manual drafting, some of which are directly linked to the areas of advantages.

To create a 2D drawing in CAD, certain key steps should be taken, such as preparing drawing units, drawing limits, layering, etc. The principles of 2D drafting,

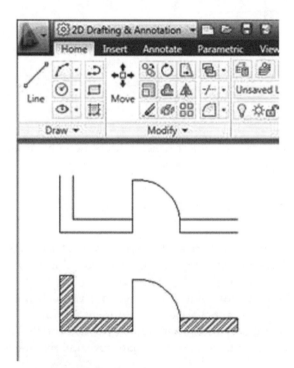

Figure 3.10 Drafting the corner of a room with a door by applying Examples I to 3 using AutoCAD LT 2011.

which are numerous, can be described as follows: suitability for the intended purpose, accuracy, legibility and neatness, economy in time and labour.

Although different 2D CAD drafting programs may have slightly differing ways of achieving the different steps listed in Section 3.2.1, the fundamental concepts are predominantly the same.

Chapter 4

3D CAD

4.1 An introduction to 3D modelling

In Chapter 3, we discussed 2D CAD, its history and attributes. Most software programs such as AutoCAD are capable of generating 3D models from 2D CAD drawings using the 'extrusion' tool, while others such as 3D Studio Max can achieve similar results by altering geometric and parametric attributes of primary shapes and solids such as circles, cylinders, squares, and cubes. (Extrusion is a common term used in 2D to describe the extension of 2D CAD 'upwards' to reflect height.) 3D models can be any real object with three dimensions that can be displayed on a personal computer using specialised software programs.

3D modelling is applicable to a wide variety of industries. For example, the video games industry uses them to create models for video games; the medical industry, to create models of such things as internal organs; the movie industry to create characters and objects for motion pictures; the science sector uses them to create highly detailed models of chemical compounds; and the architecture industry to create models of proposed buildings, landscapes and other design elements. However, the focus of this book is on buildings.

3D architectural rendering is also an important element in the modelling process, as it helps visualisation during modelling. Photorealistic rendering, which in most software programs involves ray tracing, brings the model to life and this can be achieved using complex 3D modelling and rendering software programs. Models are rendered through specific applications into two-dimensional renders of frames. Photorealistic rendering is normally desirable when architectural visualisation is required, and can also be used in many areas such as virtual simulations and 3D worlds, demos, walkthroughs, etc. This chapter aims to depict the purpose and use of 3D modelling in the AEC/ FM industry.

4.1.1 The history of 3D modelling

The art of 3D modelling in general can be traced back to late the 1800s and early 1900s. During this period, 3D clay character models were created and animated; this was referred to as claymation. Some of the first claymation stop motion movies were released with great success. Some early examples of stop motion film techniques can be seen in the *The Humpty Dumpty Circus* (1898), *Fun in a Bakery Shop* (1902), and *The Haunted Hotel* (1907). In the mid-twentieth century, 3D modelling attained something of a golden era in the movie industry: Arch Oboler released *Bwana Devil* in 1952, and Warner Bros *House of Wax* in 1953. However, in 1962 the first 3D CAD programs used algorithms to display 2D and 3D patterns.

In 1965, Charles Lang's team at Cambridge University's Computing Laboratory (which included Donald Welbourn and A.R. Forrest) began research into 3D-modelling CAD software. French researchers at Citroen, which included Paul de Casteljau, were able to significantly develop computations on surface geometries and complex 3D curves. Indeed, the work of de Casteljau and Bezier provided one of the foundations of 3D CAD software. However, MIT (Massachusetts Institute of Technology; S.A. Coons in 1967) and Cambridge University (A.R. Forrest, in 1968) were also very active in furthering research into the implementation of complex 3D curves and surface modelling using CAD software. In 1982, CATVIA version 1 was released as an add-on for 3D design and surface modelling (see Table 1.1) and in the late 1980s there was a large-scale introduction of computer- aided design systems (Björk and Laakso 2009)

Today's 3D modelling technology in the field of architecture or construction can be rendered to simulate reality for visualisation purposes. It also has been developed to include the fourth dimension (time). It is now possible to attach descriptive and quantitative attributes to various elements of any 3D architectural model (BIM). However, the BIM attribute of 3D will be discussed in Chapter 6.

Discussion

- What are the advantages and disadvantages of 2D CAD drafting over 2D (hand) drawings?
- What is the difference between 2D and 3D?

4.1.2 The purpose of 3D modelling

Three-dimensional (3D) modelling of buildings offers numerous benefits over 2D computer-aided drafting (CAD drafting) for AEC/FM professionals. On numerous occasions, 2D drawings of a proposed construction project prepared by the architect or designer have one aspect or another omitted; the term 'clash detection' is also commonly used to describe clashes of components when visualised in 3D. A well-executed 3D model makes it easier to avoid such errors and omissions.

Not only do architects gain a huge advantage in avoiding errors when putting their information into a real-world 3D environment, such advantages are also being harnessed by other industries in areas such as product development and manufacture.

The visual clarity and context coupled with the potential for reduced error has made 3D modelling a fast-growing design/presentation approach within the AEC/FM community and beyond. These models are predominantly executed using CAD software programs, which are used more frequently in developed countries. Similar to 2D (but in a more informative manner), 3D modelling can help in assisting engineers and designers in a wide variety of industries to design and manufacture products ranging from buildings, bridges, roads, aircraft, ships and cars to digital cameras, mobile phones, TVs, and clothing, etc.

Some of the key features of 3D modelling include:

- Easier visualisation processes
- Virtual simulations
- Walk-through generation
- Achieving more accurate and consistent design.

4.1.3 The advantages of 3D modelling

Modelling in 3D has numerous advantages, some of which are shared across a wide range of industries which use it. However, some advantages of 3D modelling are particular to specific industries. This section will focus on the advantages that are relevant to the AEC/FM community:

- Productivity, with the flexibility to modify designs at different levels: The majority of 3D CAD software programs enable the user to alter elements and dimensions at any stage of the modelling process without having to repeat those steps in other views. Changes made to the model are simultaneously reflected in all the available views of that particular model.
- The prototyping attribute of 3D models: 3D models are generated to represent a computer copy of the proposed real-life building or product. This makes it possible to see a building or product before it's completed.
- 3D models can also be used as marketing tools: 3D models are a tremendous marketing tool and can create additional revenue. This can be achieved by the presentation of a *3D image* on a 2D surface; a *walkthrough*, where the client is taken into a pre-recorded walk through the building in question; or a virtual-reality interactive model, where the client is not only taken into the building but can also decide where to go or what to look at.
- Increased acceptability within the AEC/FM industry: 3D modelling has become a widely accepted technology for the design of buildings and building structures, making the process of engineering design and detailing more effective and productive when compared to 2D CAD.
- Enhanced competitiveness: 3D design gives the industry a competitive edge when bidding for different forms of design/building contracts or consultancy work, as 3D visual communication of ideas can be more persuasive in winning over a client when compared to a 2D presentation.
- Design communication: 3D models are great for design reviews as they can effectively communicate the design to other members of a design team and industry professionals connected to the project.

- Error reduction: Modelling a 3D prototype before commencing production of a building or product makes it easier to spot design errors when compared to a 2D version of the same drawing. Construction and shop fabrication drawings created from the 3D model can help ensure accuracy in the final product.
- Design variety: 3D opens up a world of design that is otherwise inaccessible. Innovative building design cocepts can be generated digitally using 3D, which otherwise might not have been possible.
- Enhanced visual attribute: Modern CAD packages which are used for 3D modelling allow panning and rotations in three dimensions. This makes it possible to view a designed object from any chosen angle, which can also be from the inside looking out.
- Construction stage foresight: Some 3D packages allow the user to model and simulate a construction process, whereby construction errors not apparent at the design stage are identified before actual construction commences and can be easily rectified. This attribute helps to manage and minimise risk and wastage throughout all stages of a construction project. During the construction stage, 3D models also give the project management team the opportunity to view construction project information (Fischer et al. 2001). In the long run, 3D design drawings prepared by architectural engineers help the AEC/FM industry to build even safer and stronger building structures (Smith 2009).

4.1.4 The disadvantages of 3D modelling

There are fewer identifiable disadvantages of 3D modelling compared to 2D drafting. Some of the disadvantages are as follows:

- Time: 3D modelling takes longer to draw up than conventional 2D. Some CAD packages such as Sketchup have taken steps to simplify things. 3D forms are created in order to save time. Various attempts have been made by 3D software program developers to make it relatively easy to create 3D models. Despite these efforts, 3D modelling still requires a significant amount of time. It is important to point out that with an increase in the desired level of detail in a model, there is also an increase in the amount of time required to execute that particular model.
- Training: Modelling in 3D also requires training, which sometimes can be expensive. There are numerous CAD modelling packages, and methods of 3D object creation and modification might differ from one package to another. For example, the surface modelling approach is slightly different from solid modelling; however, a single 3D CAD program such as AutoCAD can support both approaches.
- Interoperability: The exchange of information between allied industry professionals can be reduced if editable 3D files cannot be exchanged effectively. The level of interoperability among 3D packages is limited. The key, industry-standard 3D software programs such as AutoCAD 3D, 3D Studio Max and

Sketchup are limited in terms of the number of other 3D programs they can share files with. For example, AutoCAD 3D has made it such that there is interoperability between its 3D files and Microstation DGN files. In some cases, AutoCAD 3D allows interoperability at 2D level with the exchange of 2D line data but not with 3D solid, surface and wireframe data: an example of such a program is SYCODE, where steps are being taken to address this problem in their Alibre Design. The move to address and improve the interoperability between programs is currently a lot more focused on BIM-capable 3D CAD programs using the IFC (Industry Foundation Class) standards.

4.2 3D modelling principles

Before commencing on a 3D modelling exercise in a CAD environment, there are some important principles which will ultimately lead to the production of a successful 3D model. It is also important to point out that some of these principles should be observed in a particular sequence while some can be applied at any time or even at a later stage in the drawing process. These principles include:

- *Realism:* A 3D model should be based on real-life references. When modelling a building, a 3D representation of a sash window should have taken its reference from a real-life sash window, based on its components, texture, geometry and even its mode of operation. Without such a reference to reality, a modelled sash window could end up looking like something very different from the impression the user intends to convey.
- *Scale and proportion:* The size and structure of objects and their relationship with one another should reflect proportions based on reality. It is possible, however, to have a particular type of object existing in reality at a different scale and only separated by functionality. An example might be where you are modelling a doll's house versus a real house. A doll's house could then be portrayed as a doll's house, but would be uninhabitable when placed within the context of other appropriately scaled objects such as people, trees, etc.
- *Volume:* The volume element is what gives a model its third-dimensional attribute. While the majority of objects can be broken down into primitive 3D volumes, there are other volumes which are asymmetrical and irregular. During the initial modelling phase, these 3D primitives can be reduced to polymeshes or placeholder models. In some CAD programs such as AutoCAD and Sketchup 3D volumes can be created by extrusion of 2D regular or irregular shapes along the z axis; this creates 3D volumes with a footprint of the original shape. Provision is usually made for the extrusion to be tapered or simply follow a predefined path. The sphere volume, however, is different from the other basic forms; with no straight lines, it cannot be generated by simple extrusion. The sphere pre-exists in most 3D CAD programs as 3D mesh which can be resized and deformed (see Figure 4.1).

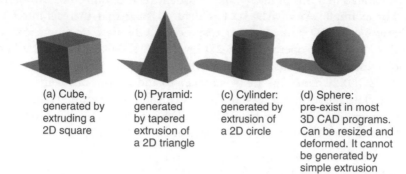

(a) Cube, generated by extruding a 2D square

(b) Pyramid: generated by tapered extrusion of a 2D triangle

(c) Cylinder: generated by extrusion of a 2D circle

(d) Sphere: pre-exist in most 3D CAD programs. Can be resized and deformed. It cannot be generated by simple extrusion

Figure 4.1 3D Primitive volumes used during 3D CAD modelling.

- *Appeal:* A 3D model should be able to communicate the design concept in terms of the elements and principles of design inherent in the building in question. If a designer incorporates elements to convey certain emotions or impressions in a building design, it is important that such information/mood is communicated in the final 3D model. The finished model should have a level of appeal. This can be achieved by good use of lines, shapes, colours, etc.
- *Exaggeration:* Sometimes the finished 3D model might, to a degree, fail to communicate the intended idea effectively. In such a circumstance, some features or rendering are exaggerated. The act of exaggeration might be applicable for 3D movies but should be discouraged for technical or parametric building models as this will communicate inaccurate design information.
- *Detail:* During a 3D-modelling exercise, paying attention to details make the objects richer, more interesting and more appealing. It is advisable to use references to discover what details might communicate and enrich your model. Catalogues of specified elements for a building can reveal significant information on desirable details which might be difficult to observe in a 2D drawing.
- *Texturing/surface treatment:* In reality the majority of surfaces have one form of texture or the other. With different surface types the chances are that you also have different texture types. In a building, the surface texture of a wall is different from that of the floor (depending on the choice of floor finish). While generating a 3D model, it is important that the surface treatment with respect to the choice of texture corresponds to the design specification issued by the designer. Texture scaling is an important part of texture editing. When a texture file is scaled too large or too small, it loses its desired identity on the model when it is applied and this error can affect the visual quality of the model. The closer the selected texture is to reality, the higher chances of a high-quality model.
- *Appropriate light and shade:* This is an important principle to adhere to during 3D modelling as inconsistent light and shade on an object relative to the light source can eliminate to a large degree the realistic quality of the model. Where 3D texture surfaces are captured under controlled illumination conditions, the

selected texture works in combination with the available lighting to enhance a model aesthetically. Most CAD programs are designed to provide accurate light and shade conditions with changes in the light source and type.

- *Coordinate systems:* Knowledge of the coordinate systems is crucial to 3D modelling because this is what separates 3D modelling from 2D drafting. Along the horizontal plane there are two main axes, x and y. The x, y axes, which are two dimensional, can be referred to a representation where 'depth' is absent. However, with the introduction of the 'z' axis depth is introduced and modelling rather than drafting is required to execute the design. The x, y, z coordinates can be rotated in multiple directions when viewing different sides of the object being modelled and can help the modeller to identify complex object faces during modelling.
- *Functionality:* In a 3D modelling exercise, it is important that all elements of an object being modelled, even roofs, facades or false accoutrements, should appear to have a purpose or function even if the function is that of decoration. This helps create the illusion of realism and a sense of interest about the object that makes it appealing.

4.2.1 Modelling

Having discussed the principles of 3D modelling it is important to note that 3D model reconstruction approaches can be either in the form of solid modelling and/or surface modelling. As the name implies, a solid model is a digital representation of a 3D object with a defined volume (see Section 4.2.1.1). A surface model, on the other hand, lays emphasis on modelling using digital surfaces to form any desired 3D objects or planes (see Section 4.2.1.2).

4.2.1.1 Solid modelling

Solid models define the volume of the object they represent (like a rock). These are more realistic, but more difficult to build. Solid modelling is distinguished from other areas in geometric modelling and computing by its emphasis on informational completeness, physical fidelity and universality (Shapiro, 2001).

Solid modelling is particularly apt for computer-aided modelling (CAM) systems which support solid modelling. This is because there are no gaps between the faces of solid models as the model has to be 'watertight'. Manufacturing software goes through painstakingly difficult mathematical functions to determine what to do if there are gaps in a surface model and to make sure the object it creates doesn't gouge (Hohler, 2000). Solid models allow for interference checking, which tests seeing if two or more objects occupy the same space.

Solid modelling is a three-dimensional modelling process in which solid characteristics of an object are built into the database so that complex internal structures can be realistically represented. (Shapiro, 2001).

Figure 4.2 Solid model showing an extruded rectangle

The real power of a solid modelling application is how it can take the solid objects and combine them together by intersecting, joining or subtracting the objects from one another to create the desired resulting shapes. Because everything in a solid model design is a 'watertight' model of the part, the solid modeller is able to know the topology of the entire model. By topology we mean that it knows what faces are adjacent to each other and which edges are tangent (Hohler, 2000).

Most solid modelling programs support geometric constraint which is a relationship of an entity to other entities. When one or more entities are 'constrained' to each other, changing any of the entities will most likely have an effect on the others. For example, in a 3D solid model, changing one dimension can affect the entire solid as they are quickly updated (in most solid modelling programs).

Some of the benefits of solid modelling include: (i) it is easy to learn/use, (ii) it possesses parametric/associative capabilities, (iii) it provides quicker creation and updating of assemblies, and (iv) it is excellent for creating functional models.

4.2.1.2 Surface modelling

These models represent the surface, e.g. the boundary of the object, not its volume (like an infinitesimally thin flexible paper). Surface models are easier to work with than solid models. Almost all visual models used in games and movies are shell models.

REGULAR SURFACE MODELLING

Surfaces can be created from curves, surface primitives or from multiple surfaces using 3D CAD software programs. This type of modelling approach is used to represent simple geometric forms with rigid dimensions such as planes, cylinders and conic surfaces. In the construction industry, the majority of buildings have surfaces which fall into this category. As a result, the majority of 3D CAD programs provide numerous tools which make it easy to generate this type of surface, such as the extrusion tool.

FREEFORM/NURB SURFACE MODELLING

NURBS (Nonuniform Rational B-Spline) is a modelling approach which involves freeform surfaces which are freely deformed and sculpted by pulling or pushing what is referred to as 'control points' (see Figure 4.3). Development of NURBS began in the 1950s by engineers who were in need of a mathematically precise representation of freeform surfaces like those used for aerodynamic cars.

A freeform surface is used in CAD and other computer graphics software to describe the skin of a 3D geometric element. Unlike regular surface, freeform surfaces do not have rigid radial dimensions due to the irregularity of the surfaces. Freeform surfaces are defined by their control points, between patches and number of patches. The control points *(known as poles)* of a surface define its shape and control polygons work together with control points for surface editing (see Figure 4.3).

SURFACE MODELLING SOFTWARE

Software programs that have surface modelling capabilities must have enough tools within the software to completely define any feature on surfaces being modelled. Such surface modelling software programs should:

- Provide enough tools to completely define any feature on surfaces being modelled.
- Have numerous functions for defining the different shapes of regular or freeform surfaces including ruled, revolved, lofted, extruded, swept, offset, filleted, blended and planar boundary.
- Support functions such as surface trimming, extending, intersecting, projecting, polygon tessellation, coordinate-system transformations and editing.
- Allow extraction of surface data such as flow curves, vectors and planes, among other functions.
- Have a set of tools for defining points, planes, vectors and splines used with surface modelling. Most surface creation functions need user inputs to define surfaces.

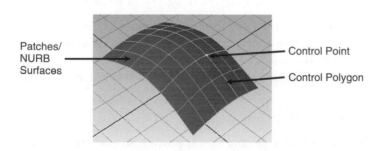

Figure 4.3 Surface model showing (i) patches/NURB surfaces, (ii) control points/poles and (iii) a control polygon.

4.3 Creating a 3D model

3D models can be created using surface or solid modelling approaches. In order to create any 3D model, it is important the purpose of the model is established. A 3D model which is required for CAM purposes will be better created as a solid model and not surface for reasons outlined in section 4.2.1.1. However, if the purpose of the model is solely for visualisation, surface models would be preferable as they are relatively lighter in terms of required computer capacity. As a result, surface modelling of a particular object are less likely to slow down your computer as compared to that of solid models of same object.

In order to create a 3D model, some key steps illustrated in section 2.2.1 which focused on 2D drafting need to be followed. Some of the relatively generic initial steps for most 3D software programs include: file creation, drawing unit assignments, drawing grid creation, and layering. Creating 3D models include: creating 3D objects from 2D shapes; navigating in 3D space; modifying and shaping 3D objects; and rendering objects.

It is important to mention that the illustrations below are generic and it is advised that the tutorials of any particular software being used should be consulted for specific instructions.

After the initial steps have been executed, the following steps outlined in Sections 4.3.1 to 4.3.4 are taken to proceed with creating a 3D model.

4.3.1 Creating 3D objects from 2D shapes

CREATION OF A BASIC 2D SHAPE

Basic 2D shapes are the shapes which act as the first step to producing the desired 3D object. Regular shapes such as circles, squares, rectangles, polygons can be drawn on a 2D plane. Most 3D software programs have standard tools which can generate these regular shapes such as circle tool, rectangle tool, etc.

FURTHER EDITING OF THE SHAPE TO A 2D FOOTPRINT OF THE DESIRED BASIC 3D FORM

These shapes need to closely resemble the footprints of the intended 3D object such as a wall etc. In some cases the desired footprints might have irregular appearances. As a result, further editing of the shape is needed. Sometimes a flexible line tool (polyline) can be used to create the irregular 2D shapes from scratch before extruding to 3D.

EXTRUSION OF A 2D SHAPE TO CREATE A 3D OBJECT

In order to create a 3D object from a 2D shape, it is possible to extrude the 2D shape along the vertical, curved path or at an angle. In this instance, a path in the form of a vertical line, curved arc, or line at an angle is created to guide the extrusion; this is the case with 3D software programs such as AutoCAD and Sketchup (see Figure 4.4).

To carry out the extrusion task described above, click on the appropriate (extrusion) tool in the software being used and follow the instructions provided by the software tutorial to complete the extrusion task. However, for software programs such as AutoCAD, you will need to click the 'extrusion' tool then following the instructions on the *command line* (this is a text window where you are allowed to type commands and view instructions in the AutoCAD software).

You will then be required to click the edited 2D shape you have created after which you will need to type the desired height and specify the path (if required) in the *command line*. On pressing 'enter' or clicking any OK button in the software being used, the 3D object is formed. The extrusion can also be carried out manually without entering dimensions. It is important to note that some software programs might have different names for the extrusion tool.

4.3.2 Navigating in 3D space

When working on a 3D project in a 3D space, there is a need to be able to observe the creation or modification of a model from different angles or views simultaneously. This is achieved by dividing the work area into multiple little work areas *(viewports)* showing different views of the model simultaneously.

There is also a need to move from one part of the model being generated to another part of it in order to either create an element or modify an element in the model. Sometimes it might be important to move closer to the element *(zoom in)* in order to accurately execute the creation or modification or move further away from the model *(zoom out)* in order to have a fuller view of what is being modelled. Sometimes you also might want to move the model from side to side without actually moving in or out *(pan)*, or even rotate the model freely in the 3D space *(orbit)*. All the above mentioned actions can be referred to as navigating in 3D space.

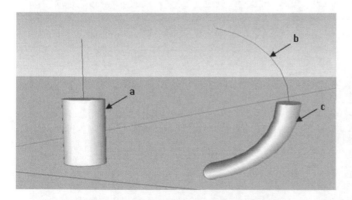

Figure 4.4 Extrusion of a circle vertically and along an arc using Sketchup:
(a) cylinder extruded vertically from a circle shape; (b) an arc which provided a path/guide for the extrusion of a circle shape;(c) extruded circle shape along the arc path 'b'.

VIEWPORT

While working on a 3D object using a 3D software program, it is important to create multiple viewports to be able to view the work area from different angles simultaneously. The term 'viewport' is used to describe the number of subdivisions within the work area. These subdivisions can come in different arrangements. For example, the work area can be single and undivided (1 viewport), or it can be divided into two halves (2 viewports), or divided into three (3 viewports) or even four (4 viewports) and so on. See an example of 4 equal viewports in Figure 4.5 (a) and 4.5 (b) (viewports 1 to 4 for Maya and AutoCAD LT software programs respectively). The maximum number of possible subdivisions and the arrangement of these subdivisions will depend on the software program being used.

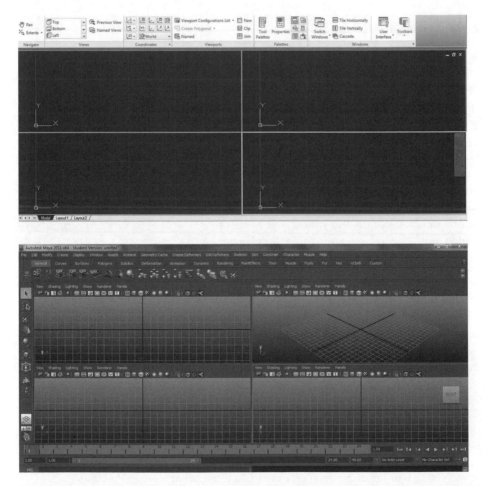

Figure 4.5 Four viewports with equal subdivision of work area. Viewports labelled from 'viewport 1 to viewport 4': (a) four viewports from AutoCADLT; (b) four viewport from Maya.

The term 'viewport' is used in AutoCAD and a few other software programs, but different names could be used to describe it depending on the program being used.

Some 3D software programs such as 3D Studio Max have four viewports available at start-up while in most cases AutoCAD has a single viewport opened at start-up. Other software programs with viewports are AutoCAD LT (for 2D) (see Figure 4.5 (a)), full AutoCAD (for 2D and 3D), and Maya (for 3D) (see Figure 4.5 (b)).

ZOOM

While working on an object in 3D it is important to simultaneously observe changes made to the object from different angles. This helps for a quicker and more accurate modelling process and multiple viewports help to achieve this.

The zoom tool is an important tool for both 2D and 3D. This tool helps enlarge the desired area of interest on the 3D object (zoom in) or reduces the object to have a wider view of the work area (zoom out). This tool is often used when attending to details or smaller parts of the object by making them large enough for editing.

PAN

While working on a particular 3D project, there are instances when the user needs to maintain the zoom level of the work area but at the same time needs to see portions of the object outside the view area. Under such circumstances, using the 'pan' tool will move the drawing area in the direction of the pan without adjusting the zoom level.

ORBIT

The 'orbit' tool is a 3D viewing tool which is referred to as such in AutoCAD. It represents your orientation in 3D space. By moving the orbit tool, you move the entire model along the x, y and z axes. This can be used to view the model during the modelling exercise. However, other 3D software programs have different names for this tool (refer to the tutorial manual of the software program being used).

4.3.3 Modifying and shaping 3D objects

A 3D object can be edited or modified until the final desired object is achieved. For example a cube in 3D can be modified and shaped to a pyramid. Methods of editing an original will depend on the nature of the 3D object: is it a solid model or a surface model?

A 3D object created as a result of solid modelling is modified slightly differently to those generated using NURBS. A NURBS surface model made up of patches can be modified by adjusting the control points (see Figure 4.3). A solid model can be modified primarily by subtraction or addition of other 3D objects. A solid model can also be modified by slicing off unwanted parts of the model.

3D solid objects can be modified by the addition and subtraction of other objects to create the desired final object.

To carry out a subtraction exercise, the primary object (the object from which something is to be subtracted) is identified. A secondary object is created with the geometry of the part to be inserted identical to the geometry of the hollow intended to be created on the primary object.

The secondary object is then placed in the position on the primary object in the area where the hollow is to be made. The appropriate subtraction command is then given depending on the software program being used. The secondary object then disappears, leaving a hollow on the primary object identical in geometry to the part of the secondary object inserted. This method can be used to create windows for example.

To carry out an addition exercise, the same procedure is followed for both the primary and secondary objects. However, this time attention is paid to the geometry of the secondary object that is to be added. During the addition process both the primary and secondary objects are joined.

Some 3D software programs make provisions for a solid model to be sliced. This task is similar to using a knife to slice through butter. The task of slicing takes place digitally and the software allows you to (i) create your knife element or 'cutting edge', (ii) position the cutting edge in the desired position on the object to be sliced, and (ii) execute the slicing task by selecting the part of the element that should remain or the part that needs to be removed (this depends on the software program).

It is important to point out that the type of command, sequence of commands or entire editing approach might differ with different programs, so it is important to consult the tutorial of the 3D software program being used.

4.3.4 Rendering objects

Rendering of 3D models helps viewers appreciate what the model represents, when colours, textures and lighting on the object are used appropriately. The rendering of modelled objects is key if visualisation is one of the intended objectives of the model.

There are different levels of rendering and the type of rendering adopted will depend on the purpose and target viewers of the model. Simple rendering is basic, less realistic, less demanding on the computer and quicker to execute. However, photo-realistic or more complex rendering is more informative, closer to reality and takes longer to execute. Sometimes wireframe rendering can be used and this is where the model is reduced to a series of lines or wires.

3D software programs make provision within the software for the user to render the model after creating it. However, some 3D software programs make provision for the use of a different rendering software programs to handle the rendering task. The rendering software can be embedded into the parent 3D software as a 'plug-in' or exist as separate software, a *'stand alone'*. For example ArchiCAD as the parent 3D software uses Atlantis as the rendering software for its models, while Sketchup as the parent 3D software uses Podium as the rendering software. The list is endless.

Rendering in 3D is a simple process but inputting settings such as lighting, shaows, resolution, etc., can be relatively complex and this can vary between rendering software programs. In most 3D software programs, the desired rendering resolution can affect the rendering time from a matter of minutes to a matter of days. However, high speed computers can provide some reduction in rendering time.

To achieve photorealistic rendering, it is important that the modeller is competent with regards to the software program being used. The model should be ready before the commencement of rendering. Select materials from the material library provided in the software or online.

Scale materials appropriately such that patterns on the materials are as close to reality as possible in terms of colour and dimensions. Ensure that reflection and transparency values have been appropriately assigned. It is important to note that opaque objects like wooden floors have a level of reflection and hence this should be taken into account while rendering.

While applying the lighting element to the object to be rendered, a decision is made as to whether day lighting or artificial lighting is to be used. Values are entered for the intensity of the lights, which in most cases are updated in the work area. If artificial lights are used, light fixtures should be positioned appropriately. While assigning values for artificial lighting, *'hot spots'* (portions of very high illumination on a surface) should be avoided.

The camera positions and target views should be identified in order to render the desired area of the object. When this is done, the rendering process can commence. The illustration in this section is generic as some software programs can be more complex. Please refer to relevant tutorials.

WIREFRAME RENDERING

During the modelling exercise, it is possible to render in wireframe as opposed to photorealistic approach. A wireframe model is the result of a rendering process which creates a visual presentation of a 3D object as being defined by multiple lines or curves. It is created by connecting an object's constituent vertices using these straight lines or curves.

Wireframe rendering of a 3D model allows the visualisation of its underlying design structure. The wireframe model in some instances can appear complex if there are multiple underlying structures within the model. This can sometimes make the complex wire network difficult to articulate. Curved objects are more likely to generate more lines or curves (cylinders and spheres) than objects with linear edges or planes (cubes and pyramids) (see Figure 4.6).

The wireframe model, however, has the advantage of being light in the sense that less computer processing power is required to manipulate the model in comparison to the other rendering approaches.

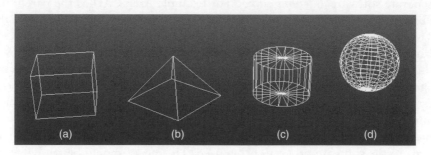

Figure 4.6 Wireframe rendering of (a) cube, (b) pyramid, (c), cylinder, and (d) sphere.

4.4 3D modelling practical examples

Some examples of 3D modelling are demonstrated in the following sections. Solid and surface modelling approaches can be employed; however, this depends on the purpose of the model. Hybrid solid and surface modelling can also be used for relatively complex models.

4.4.1 Example 1: Surface modelling

Different 3D software programs provide different ways of creating surface models. In previous versions of AutoCAD, surface models can be created by drawing a polyline and assigning thickness 0 to the polyline. The value assigned results in a form of extrusion of that line along the vertical axis.

Other software programs such as Sketchup provides an interactive and flexible approach. Shape tools such as the 'rectangle' or 'circle' tools can be used to create an initial 2D shape. A *'push/pull'* tool is then used to drag the 2D shape vertically to create a 3D surface model in a form of extrusion. Values such as height can be assigned rather than the vertical drag.

4.4.2 Example 2: 3D solid modelling

As solid modelling involves defining an object with geometric mass, solid modelling programs usually create models by creating a base or primary solid and adding to or subtracting from it. They can be modified using features such as extrudes, extrude cuts, revolves, radii, chamfers, etc. (Finkle, 2011). See Section 4.3.3.

4.4.3 Example 3: Modelling a simple house

In this example, Sketchup 7 was used to produce a simple 3D model of the outer shell of a house. For more complex 3D models, a detailed floor plan layout generated in the 3D program being used or in another program forms the footprint for the model.

Step 1: The 2D rectangular boundary '*a*', which represents the footprint of the intended 3D model, is generated in the Sketchup software (see Figure 4.5 (a)). The shape of the footprint generated must not be a rectangle as this will depend on the footprint and overall geometry of the intended model. However, the polyline must form a closed boundary, i.e. the start and end point of the line used to create the shape of the footprint must meet.

The perimeter walls are generated by drawing an inner rectangle '*b*' (see Figure 4.5 (b)). The thickness of the wall is determined by the user or designer. At this stage the steps are carried out in a 2D plane, i.e. along the x and y axes.

Step 2: With the inner rectangle which marks out the perimeter wall in place, the inner segment (rectangle footprint) '*b*' and outer segment (perimeter wall footprint) '*c*' are now separate entities.

With this separation in place the segment (perimeter wall footprint) '*c*' is then pulled up to the desired height using the '*push/pull*' tool (see Figure 4.6 (a)). This height can be prescribed by entering the desired values in the relevant box within the Sketchup program interface.

A similar process is repeated without the inner rectangle to create a larger rectangle '*l*' for fascia/roof representations (see Figure 4.6 (b)). The rectangle plane, however, was created with slightly larger length and width dimensions to make provision for the eaves, and pushed up to create the fascia '*f*'.

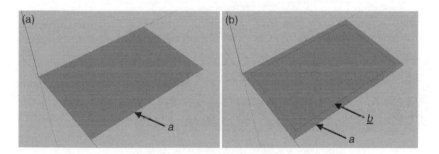

Figure 4.7 (a) 2D rectangular boundary (perimeter wall footprint); (b) inner rectangle showing wall thickness.

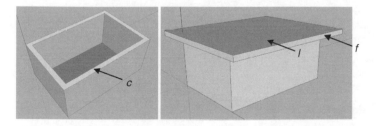

Figure 4.8 (a) Rectangular boundary/perimeter wall extruded/pulled up; (b) larger rectangle for fascia/roof representations added on top of wall.

Step 3:   The next step is to create a gable roof. In order to achieve this, it was important that the upper rectangle '*l*' be divided in the middle to provide a handle with which to raise the gable roof. The gable roof pitch was raised from the handle using the Sketchup '*move*' tool.

With the roof in place, a similar exercise as 'step 1' was carried out to mark out the door '*d*' and window '*w*' openings. Rather than pull, the '*push pull*' tool was used this time to push the marked door rectangle and the marked window rectangle, which were then separate segments (see Figure 4.7 (a)). Openings were hence created on the wall to provide for two windows and a door (see Figure 4.7 (b)).

Attention was then given to the door and window frame representations by using similar approaches to those described above. However, to create the frames, the exercise was carried out on a smaller scale.

Step 4:   On completion of the modelling stage, the final step involved material assignment and rendering.

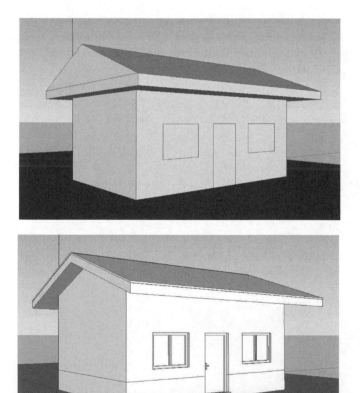

Figure 4.9 (a) 3D model with walls and roof in place with outline of door and windows; (b) further details of roof door and window added.

The rendering exercise has been generically described in Section 4.3.4. Materials were selected from the material library provided by the Sketchup software. The materials used (brick, wood and roof tiles) were scaled appropriately with patterns as close to reality as possible in terms of colour and dimensions and then applied (see Figure 4.10 (a)).

Basic lighting was applied to the 3D image. The finished model was then rendered using Sketchup plug-in render software (Podium) to create a 3D image (see Figure 4.10 (b)).

The above example has been simplified. In some programs, and for more complex buildings, further steps not mentioned above may be needed. However, the above example provides an overview on the key steps required for 3D modelling.

4.5 Summary

In this chapter, we have established that three-dimensional (3D) modelling of buildings offers numerous benefits over 2D computer-aided drafting (CAD drafting) for AEC/FM professionals.

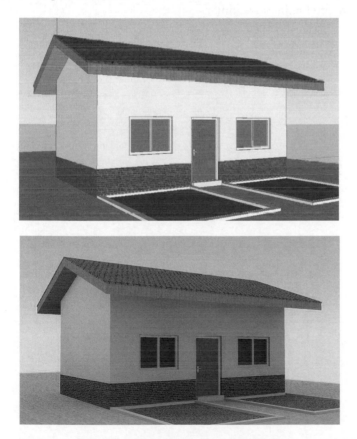

Figure 4.10 (a) Complete 3D model with materials assigned but without rendering; (b) rendered complete 3D model.

Modelling in 3D has numerous advantages, some of which are shared across a wide range of industries which use 3D to reduce errors and communicate designs with enhanced visual attributes, etc. However, some advantages of 3D modelling are peculiar to specific industries, such as the construction industry where foresight, though appreciated by stakeholders, is critical to contractors. There are fewer identifiable disadvantages of 3D modelling compared to 2D drafting.

There are certain principles to consider while modelling in 3D. These principles should be observed in a particular sequence while some can be applied at any time or even at a later stage in the drawing process. These principles involve realism, scale, volume, etc. (see Section 4.2).

The modelling approach can either be solid modelling or surface modelling. Either one of them have areas where they can be best applied. Irrespective of the modelling approach, solid or surface, there are fundamental steps to take while modelling in 3D, which have been discussed in Section 4.3.

It is also important to remember that different 3D CAD drafting programs might have slightly different ways of achieving the different steps listed in Section 4.3 but the fundamental concepts are predominantly similar.

Chapter 5

BIM (Building Information Modelling)

5.1 An Introduction to BIM

In the previous chapter, the concept of 3D CAD modelling has been discussed. However, modelling can be taken a step further with the introduction of the possibility of adding realistic information to the elements within the model. Using a building as an example, if we apply additional information such as the types of material, the costs, etc. to the different elements within the 3D CAD model, the generated model is then capable of replicating a real-life building scenario. The building prototype is replicated in 3D CAD not only visually, but with a high level of accuracy, containing actual building information within the model. This is more commonly referred to as Building Information Modelling (BIM). BIM *also* refers to constructing and managing a building information model over time. Moreover, there are various BIM software packages available which enable professionals to virtually build a facility before it is ever physically constructed.

BIM can be considered an ambiguous term (Aranda-Mena et al., 2008). There are definitions that present BIM as a 'technology' or as a 'process', whilst others still hold the view that BIM is not about technology or process but about the building (Bedrick, 2005). There are also definitions that present BIM not just as a process or technology but as a strategy. However, simpler definitions present BIM as a digital representation of the physical and functional characteristics of a facility.

Whatever definition is given to BIM, the function of BIM involves the detailed and complete replication of a building in a digital environment and this is the basis of its value to the AEC/FM industry from which other benefits are derived. The goal of BIM is for architects, construction firms and owners is to collaborate and use integrated digital and database tools throughout the procurement process of a building, which includes its commissioning, design, build, and maintenance cycles. BIM modelling as an integrated design method has brought significant advances to the design and delivery of new construction projects (Attar et al., 2010).

The key differences between a 3D CAD model and a BIM model are an increased level of modelling accuracy and the introduction of a database which consist of attributes linked to the model and its elements. Such data sets include the building geometry and spatial data as well as the properties and quantities of the components used in the design.

By adopting BIM, architects, engineers, contractor's operators and owners can easily create coordinated digital design information and documentation (Boutwell, 2008) and use that information to more accurately visualise, simulate and analyse the performance and appearance of buildings (Lenard, 2010). In comparison, designers using only 2D or 3D CAD applications require numerous specification sheets in order to convey all the required information pertaining to the project.

An important factor in the success of a BIM model is its ability to encapsulate, organise and relate information which can be understood by its user and also by other software programs. The information within the model must be at a highly detailed level, relating, for example, a window to its frame or even a nut to a bolt, while maintaining the relationship between the detailed level and the overall view. However, it is also important to point out that BIM is not all 3D. Within a BIM model, there still are numerous components that may be represented as flat 2D objects, or even just listed as data in a schedule.

Like the adoption of 2D drafting and 3D CAD modelling, there are advantages as well as disadvantages involved. It is important to point out that it is widely assumed that the advantages of BIM modelling far outweigh its disadvantages.

5.1.1 The history of BIM

The term BIM was coined by Charles M. Eastman at Georgia Tech and has been around as a concept since the 1970s. However, the technological advancement of computers in terms of their processing capabilities, at speeds which were not attainable in the 1970s, is a key factor in the implementation of the BIM concept, and software developers have responded to the demand by practitioners (Yessios, 2004).

In 1986, Graphisoft introduced their first 'Virtual Building Solution', known as ArchiCAD (Kmethy, 2008). The software was revolutionary as it allowed designers to create 3D representations with the ability to store large amounts of information within the 3D model. As BIM is based on parametric modelling, the geometric consistency and integrity of the building model is maintained in spite of any changes or modifications that are made to it.

More recently, BIM has risen in popularity within the AEC/FM industry. McGraw-Hill Construction's latest SmartMarket Report for 2010 on the business value of BIM reveals that 49 percent of respondents report they are using BIM tools, which is a 75 percent increase over the 28 percent BIM adoption rate measured in 2007 (McGraw Hill Construction, 2011). This shows a remarkable increase in the adoption of BIM within the construction industry as its numerous benefits are being identified.

With growing acceptance among the AEC/FM industry, BIM users are now able to represent their designs in a single 3D database which can be continually updated as changes are made or errors are detected during the design delivery stages and even into the facilities management stage of the building in question. Accuracy, time and cost savings are potentially realised. With the rise in BIM adoption, software manufacturers have been developing programs expected to satisfy the BIM requirements. Some of the most common software programs include:

- *Autodesk Architectural Desktop (ADT)* provides a transitional approach to BIM, as an intermediate step from CAD. ADT creates its building model as a loosely coupled collection of drawings, each representing a portion of the complete BIM. These drawings are aggregated through various mechanisms to generate additional views of the building, reports and schedules as though there was a single BIM at the centre. One problem with this approach is complexity in managing this loosely coupled collection of drawings and the opportunity for errors if the user manipulates the individual files outside the drawing management capabilities provided in ADT.

- *Autodesk REVIT* is perhaps the most literal interpretation of a single BIM as a central project database. The strength of this approach is the ability to coordinate every building element in one database, thus providing users with the ability to immediately see the results of any design revisions made in the model, have them reflected in the associated views (drawings), as well as to detect any coordination issues. Previous versions of REVIT as a proprietary data model did not support IFC (industry foundation classes: a neutral and open specification that is not controlled by a single vendor or group of vendors) import/export. However, current versions are made to support IFC.

- *Bentley Systems* interprets BIM differently as an integrated project model which comprises a family of application modules that include Bentley Architecture (which is also still sold in some international markets under its original Microstation Triforma name), Bentley Structures, Bentley HVAC, etc. Bentley describes this approach as an evolutionary path that allows its Microstation users to migrate work practices that have their origins based on using CAD. Access to project data is provided with DWG and IFC file formats both being supported. However, the highest levels of interoperability are only achieved when the entire family of Bentley products are deployed on a project.

- *Graphisoft's* approach to BIM is to create a virtual building model, meaning their ArchiCAD application is viewed as one of many satellite applications orbiting a virtual building model rather than being seen as the central repository for the entire model. In addition to ArchiCAD being conceived as a BIM system from its inception over 20 years ago, Graphisoft is now working with a consortia of application partners to deploy EPM Technology's IFC-based model server as a virtual building repository, possibly the most innovative technical approach to the future of BIM.

5.1.2 The purpose of BIM

BIM is used to produce a digital prototype of a building in order to understand its behavior before construction. As a result, BIM bridges communication between design and construction teams.

BIM is able to achieve many improvements by modelling virtual prototypes of the actual parts and pieces being used to build any particular building. This is a substantial shift from the traditional computer-aided 2D drafting method of drawing with vector file-based lines that combine to represent objects and 3D representations of the building which are limited to visual aesthetics.

In order to achieve a standard format for viewing BIM models by all members of the team involved in a project, BIM is often associated with IFCs (Industry Foundation Classes) and aecXML, which are data structures for representing information used in BIM. IFCs were developed by BuildingSMART (International Alliance for Interoperability). Other existing data structures are proprietary, and many have been developed by CAD firms that are now incorporating BIM into their software.

Freeman (2009)'s definition of BIM as the natural progression of the use of computer-aided design (CAD) tools to combine graphic objects with parametric dimensions in a way that simulates actual construction results even before ground-breaking suggests that the BIM model can also serve a 3D graphic/aesthetic communication purpose. The progression has been from 2D to 3D, 4D, and BIM. However, while 3D models make valuable contributions to communications, not all 3D models qualify as BIM models since a 3D geometric representation is only part of the BIM concept (GSA, 2010).

BIM is an integrated digital description of a building and its site comprises objects described by accurate 3D geometry, with attributes that define the detailed description of the building part or element, and its relationships to other objects (Mitchell and Schevers, 2006). The virtual world potential achieved by adopting the immersive technology of the BIM model creates the opportunity for a stakeholder to gain a total and immersive understanding of the building (Okeil, 2010). It is able to bring together the different threads of information used in construction into a single operating environment, thus reducing, and often eliminating, the need for the many different types of paper documents (Froese, 2008). It improves collaboration and aids faster question-and-answer turnaround time between project team members (Kymell, 2008). BIM also can be used as an information framework for storing and retrieving FM related data (Freeman, 2009).

In March 2008, the US General Services Administration's (GSA) Public Buildings Service (PBS) signed an agreement with three international real estate organisations to support open standards for BIM software and systems. It has also mandated that 'every new facility and major modification project should utilise a BIM model for spatial validation' (Boutwell, 2008, cited in Gillard et al. 2008).

5.2 BIM applications within the AEC/FM industry

Efficiency in the building delivery process among the AEC/FM community can be improved with the elimination or reduction of operational bottlenecks such as errors and omissions that can occur with paper documentation. Such errors can lead to cost overrun and project delays.

5.2.1 BIM for designers (architects and engineers)

Architectural services in terms of building procurement can be said to consist of the actual design of the building and its analysis. Architects work with engineers during the design phase of a building to ensure that structural solutions are provided for design concepts.

During the building procurement process, the design needs to be communicated to the client and other members of the design team such as the engineers etc. This communication is important as it enables the designer to arrive at a suitable design solution which satisfies the client as the major stakeholder. For this reason, visualisation is an important aspect of the design process. Some visualisation approaches can include: *3D design, design review, detailing and documentation, rendering and graphical representation*. Also included are *virtual reality and augmented reality* (not discussed in this book).

A building during the design process is also expected to take certain issues into consideration in order to be successful. Such issues involve: *quantification and estimation, fire analysis, code compliance checks, spatial coordination and clash detection, sustainability and carbon footprint*. Also included are *site analysis* and *lifecycle costing* (not discussed in this book).

BIM as a tool can be used by the designer (architect or engineer) to carry out the design visualisation and analytical functions listed in this section. However, the various capabilities to carry out these functions could differ from one BIM model to another and in some cases might require collaborative software programs.

5.2.1.1 Visualisation

3D DESIGN

3D and BIM technologies represent separate but synergistic ways which the designer can use to achieve design goals. BIM modelling is not 3D modelling. However, BIM models are usually represented in 3D but contain more information about the building than a 3D model. During the conceptual design stage of the building, where the massing, placement and general appearance of the building is decided, the BIM model can be used to generate 3D objects which can be modified and moved around following design decisions. After this has been satisfied, more details can then be added until the design is complete.

DESIGN REVIEW

This involves a panel of design professionals reviewing a design to ensure good, cost effective and quality design solutions. Visualisation is important for the design review to be achieved. The design professionals involved with the design review might have used different BIM software, but in order to exchange BIM information, the BIM tools used should interconnect (interoperability). There should be a collaborative production environment which is critical to ensuring successful project resolution and greater efficiencies. There are design review tools that allow different BIM platforms to communicate seamlesly, ensuring that the same goals are achieved. These tools generates design review checklists which can be used to communicate to designers issues their designs should address. Without a design review there would be no BIM coordination process as we know it today.

DETAILING AND DOCUMENTATION

During a design exercise, while transitioning building elements from early concept to the detailing stage, it is important to detail accurately as a wrong detail could be expensive as well as result in delayed project delivery. The BIM model allows the designer to achieve a very high level of detailing which extends from the physical character and dimensional detail (which can be visual) to the specification and manufacturing detail of any particular building element, and this is then stored as a text document in the BIM database.

RENDERING AND GRAPHICAL REPRESENTATION

On completing the concept stage of the building, the ideas needs to be conveyed to the client in a way that can be easily understood. A variety of BIM software programs have a relatively good rendering capability. This means that while developing the BIM model to achieve the desired concept, the designer can proceed to render the model by applying the traditional rendering techniques found in the regular 3D software (see Chapter 4, Section 4.3.4).

5.2.1.2 Analysis

QUANTIFICATION AND ESTIMATION

While designing a building is the responsibility of the architect, cost-estimating is the responsibility of the estimator/quantity surveyor. Model-based quantification and estimation is made possible in a BIM model by computable building information. A BIM model stores the type, dimensions and number of all building elements present within the model and this is updated automatically. As a result of the accuracy of the model information, this BIM capability makes it possible to extract material takeoffs from the underlying model.

FIRE ANALYSIS

The BIM software can be used to analyse fire safety issues without having any specialised software. For the majority of building regulations all over the world, the issue of compartmentation and fire resistance of selected building materials play a key part in addressing passive fire safety issues.

The BIM model has a database which lists all types of materials and their fire properties (if such information has been included). The 2D and 3D visualisation capability of the BIM software makes it possible for the compartmentation issues to be identified and addressed during the design stage. For example, the fire properties of a wall can be retrieved by clicking the compartment wall in question and then retrieving the properties of the highlighted wall (the execution of this task can vary with different BIM software). Active fire safety issues such as sprinklers can also be analysed from the BIM model with respect to types and positions when checked against regulatory requirements.

CODE COMPLIANCE CHECKS

Validating compliance with a building code is easier with BIM. Automatic checks on code compliance in a BIM model can be carried out. This is because at the core of BIM lies a digital database in which objects, spaces and facility databases are identified and stored (Madsen, 2008). This database is checked by the BIM model or the BIM model is checked for certain code compliance parameters by separate software. One example is through the use of BIM and an autocode check tool such as 'buildingSmart', whereby users can automatically check compliance with international energy conservation codes. Other such codes might include seismic codes, international fire codes, international mechanical codes, etc.

CLASH DETECTION

Bringing together BIM design models created by different professionals and by different types of BIM applications can be a challenge. One model might be from the architect, another from the structural consultant and yet another from the MEP (mechanical, electrical and plumbing) consultant, and there might be slight discrepancies with resulting clashes. Clashes might be *'hard clashes'* (where two objects are occupying the same physical space), *'soft clashes'* (where an object occupies a space which is required for another object to function), or *'time dependent clashes'* (something that is happening in a sequence that can't happen in reality).

Some BIM software programs are designed to detect these clashes, while in other cases separate clash detection software programs like 'Navisworks' help to detect such clashes.

SUSTAINABILITY

Sustainable design attempts to reduce the negative impact on the environment by adopting environmentally friendly design and construction practices. BIM software programs are useful to designers when it comes to addressing sustainable issues such as energy.

BIM software can help in revealing energy efficiency attributes in multiple design alternatives. This can be achieved by the calculation of the southern orientation for day lighting and views for each alternative. The findings could, in turn, help to determine the position and sizes of south-facing windows in order to make the best use of the sun. When using BIM software, the optimum angle for solar panels can also be determined, based on the sun's path and the sun's angle at different times of the day and year. Though energy analysis does not require BIM, using the model greatly facilitates it.

5.2.2 BIM for contractors

Construction sites can be quite busy, and for some buildings some of the sub-elements are either manufactured offsite or onsite. The execution of both onsite construction and offsite manufacturing can benefit from BIM.

5.2.2.1 Onsite construction

CONSTRUCTION SPECIFICATIONS

Building specifications, which are usually produced by the designer, are laid out in a document which describes in detail, the types and sources of the different building elements which make up a particular building. This information can be stored in a database by the BIM software in such a way that it can be referenced, displayed and even quantified. Some BIM software programs come with predefined elements with some levels of specification which can be adopted, improved upon or altered by the designer.

PLANNING AND SCHEDULING

The aspect of planning and scheduling of any building project is vital to save time and cost. The BIM model's extensive database brings under one umbrella all of the building information required for planning in its construction phase. However, in addition to the available information there is the *'time element'* which is vital for scheduling. A 4D (3D + time) planning and constructability tool allows a BIM model created in software programs such as REVIT to be integrated with either MS Project or Primavera software programs. Different building construction tasks are then optimally organised along a timeline. This will be discussed in greater detail in Chapter 6.

TRADE COORDINATION

During the construction stage different professionals and trades are usually involved in the project delivery process. For these trades to function effectively, and to avoid time overlaps for those trades which cannot function simultaneously, it is important for the contractor to understand the building in terms of what needs to be done and in what order. The BIM model that is the virtual copy of the complete building will help the contractor identify the different trade tasks required and then determine what needs to be done first and what can and cannot be carried out simultaneously. This can be achieved through a virtual construction exercise using the BIM software program.

Construction bidding is the process of submitting a document which stipulates the amount of money the bidder will charge to undertake any construction works. Such documents are based on accurate information about the building provided by the designer. The BIM model database provides the bidder with information on various elements of the building, types and the quantities of materials etc. The BIM software program is also able to demonstrate the virtual construction exercise, which can help to determine the type and amount of labour involved. This information will then aid the bidder in calculating the amount of money which will be contained in the bid document.

5.2.2.2 Offsite manufacturing

STEEL FABRICATION AND PRE-CAST CONCRETE

To fabricate steel, a hot rolling manufacturing process is used to create steel members which are then purchased and cut by fabricators to create steel beams using shop drawings. Shop drawings are either manually or digitally generated. BIM software can then be used to create the shop drawings/models with a database which describes the attributes of the steel. The BIM hence enables digital design-to-fabrication work-flows for structural steel designers.

5.2.3 BIM for facility managers

Facilities management is the total management of all services that support the core business of an organisation, which includes its buildings, and the person who carries out the facilities management task is the facilities manager.

5.2.3.1 Management

The management of buildings by the facilities manager can be complex. Mechanical and electrical equipment such as that used for ventilation, lighting, power systems, fire systems and security systems in a building needs to be well monitored and maintained for building operational efficiency. The equipment used in these systems can be monitored using BMS (building management systems). Moreover, involving a BIM model of the facility will provide a context in which the benefits of BMS will be enhanced.

Some of the tasks involved in facilities management that can benefit from BIM include building maintenance, building tracking, space management and keeping of accurate building records.

BUILDING MAINTENANCE

In order to effectively maintain a building, information about the building fabric and operations that are carried out within that particular building is crucial. During the operational period of a building the various building components and machinery which aid its operation will require replacement from time to time. In the event of a building problem such as a water leakage, the BIM model as a virtual replica of the facility will not only visually represent the leakage but also show if there are plumbing works, pipe connections, etc. near the leakage in both visual and data formats. This information will assist the facilities manager to easily identify and solve the problem, hence saving time and cost.

SPACE MANAGEMENT

During the operational lifecycle of a facility, space allocation could be dynamic which means that different spaces could have different functions depending on occupancy requirements. The BIM model which has all the spaces within a building fully represented, and attributes like room sizes, geometries and wall types stored in a database, could be an important space management tool. With the BIM model, the facilities manager is able to make informed decisions on space prioritisation and allocation.

BUILDING RECORDS

It is important to monitor the overall performance of buildings during their operational lifecycle. Building records are information about the building from construction through to its operational stage. Such information can be stored in a BIM model and will provide vital information about the building in terms of overall cost, energy performance, operational efficiency, etc.

5.2.3.2 Simulation

Simulation in this section is described as the act of digitally imitating real-life behaviour of some situation or process. Simulation capabilities have become a key technology in numerous BIM modelling programs. Simulations can be carried out on different aspects of the BIM model of a building to investigate certain attributes of the facility or see how it will perform under certain conditions in terms of energy, occupancy traffics, etc.

ENERGY CONSUMPTION ANALYSIS

A BIM model can be used to carry out a time-based simulation of the energy use of a building and overall energy cost. Lifecycle estimate of the energy use and cost can also be calculated using information from a BIM model.

OCCUPANCY TRACKING

Information on movement within a building by its occupants can help facility managers and designers to identify how the facility is responding to the traffic needs of its users. In some cases the choices made by the facilities managers on space allocation can create bottlenecks at points within the building during peak traffic periods. Occupancy tracking executed by running simulations on the BIM model can help to identify potential problem areas and create an early warning system for the facilities manager who is then able to act to eliminate such potential problem. (See tutorials of adopted BIM modelling software for steps on how to carry out above simulation.)

5.3 The advantages and disadvantages of BIM

It is widely said that BIM's advantages far outweigh its disadvantages, and because BIM is fairly 'new' it will take time to realise its full advantages.

5.3.1 The advantages of BIM

Some identified BIM advantages are:

- Increases design productivity and quality: Design productivity is increased as a result of easy retrieval of information, which also helps in cost and time savings.
- Early modification: Another advantage is the ability to make changes easily, even late in the project. This means that late changes to a design may be done with much less effort than previously, and that more options can be explored during the early design process. In summary, design issues can be addressed and modified earlier in the process as a result of improved visualisation, saving time and reducing cost.
- Construction planning: BIM models allow for improved construction planning and increased coordination of construction documents. In the BIM process all of the information comes from a single file, so it is always up to date and co-ordinated. Coordination means that when you change a door in one place, that door will be updated in every other place. In contrast, when using a traditional CAD system, the person drawing must remember every place where that door occurs and go back into the many drawing files and manually change each door in each view. Also, every trade and profession in the construction industry has the ability to speak the same language regarding the model, from the lead architect responsible for the design to the contractor and project's insurer.
- Problem identification: Helps identify problem areas within the design which would not have been easy to identify using 2D or 3D CAD.
- Interoperability: Adopting the BIM technique will also enable interoperability amongst industry professionals as the Industry Foundation Classes (IFC) creates the desired platform for facility managers to share digital datasets (Gillard et al., 2008). The importance of the BIM tool with its interoperable platform and dataset is reflected in the National Institute of Standards and Technology's study

(NIST 2004), which estimated that the cost of inadequate interoperability to the US capital facilities industry was approximately $15.8 billion in 2004.

- Additional information accommodation: BIM is capable of embedding and linking vital information such as the names of vendors of specific materials, the location of details and quantities required for estimation and tendering.
- Design flexibility: BIM allows for more flexibility from the design of the project to the actual construction. It enables designers, contractors and owners to work through the model together to implement changes easily and efficiently.
- Simultaneous modification: Modifications are made easily in each phase of the model with BIM. Instead of going back to the drawing board, you simply change it in one place and the changes flow through all of the affected details. When a dimension or property of a component is altered, it is recognised by the model, and it modifies every database that deals with that element. This attribute helps avoid any future conflict that would normally slow the building process.
- Document extraction: BIM has the technology to produce several user-friendly documents needed during the course of a project. A design project can contain over 10,000 pages of documents. However, with BIM you have a model from which any of these documents can be produced. This feature eliminates field or shop drawings by having all parties work within the shared model. Two- and three-dimensional PDF files can also be generated to give the owner or employees enhanced visualisations of the design process.
- Data extraction for analysis: BIM provides specialised analysis tools to extract data from the design process and perform valuable analysis. The different professions in construction, such as the architects, engineers, etc., use this capability for different tasks. Any category of data needed can be obtained from the work done in a BIM process.
- 3D views: One of the largest advantages is the ability to easily produce a number of 3D views that are more easily understood than traditional plan and elevation views. This is particularly useful for unique or complex designs.
- Material takeoffs: The Building Information Model allows us to request certain information from the database in order to have an idea of the quantities of certain types of materials required for the design. Such information can include: 'How many cubic metres of concrete are needed for the foundations?' or 'How many square metres of this type of timber are required?'
- Accuracy: Unlike 3D CAD models, where accuracy is not of prime importance because of the models' main function of enabling visual communication, the BIM model has to be accurate as every architectural and engineering representation has to fit within an integrated data environment. As a result there are high accuracy levels involved in BIM representations.
- Faster and more effective processes: Information is more easily shared, can be value-added and reused.
- Better design: building proposals can be rigorously analysed, simulations can be performed quickly and performance benchmarked, enabling improved and innovative solutions.

- Controlled whole-life costs and environmental data: The BIM model can also be used to understand and predict the environmental performance of a building and its lifecycle costs during the management period of the facility.
- Automation assembly: BIM data can be exploited during facilities management, ensuring that procurement decisions are made on the basis of whole-life costs and cultural fit, and not solely on short-term financial criteria. Proposals are better understood through accurate visualisation of lifecycle data requirements. Design, construction and operational information can be used in facilities management of the building.

5.3.2 The disadvantages of BIM

BIM adoption also has some disadvantages, although these may change as more and more AEC/FM practitioners adopt the modelling technique:

- Manpower cost: The popularity of BIM has grown tremendously in the past decade, but so too has the demand for well-trained designers and construction managers with proficiency in the use of BIM technology. There is a large upfront time and manpower cost involved in building a 3D model. Adequate training is needed in different areas, and levels of BIM expertise can vary. This is especially an inconvenience when the onsite field workers are not as familiar and aware of BIM as the construction managers would like and the managers cannot afford the cost of training.
- Errors in accuracy: In BIM, since the model is the core aspect of the project, just one error in precision by any of the model contributors can be very costly. Unlike the traditional 2D project method, there are several different sets of plans that can be used to check one against the other and prevent such mistakes. With BIM, the plans are generated from the model, so they all reflect the same data, making it harder to catch small miscalculations that can lead to bigger problems.
- Software cost: In BIM technologies, software costs and required hardware upgrades are costly and it takes a lot of time to implement them into an existing process. The problem here is that because of the growing importance of BIM, the software vendors could price out SMEs (small and medium sized enterprises) from easy acquisition and implementation of the BIM program within their organisations.

5.4 BIM modelling principles

BIM DATA REQUIREMENT

Building model data should comply with interoperable file format. Intending users of building model data in any particular file format should ensure that the most appropriate data format is selected and will enhance data sharing process with other members of the building team involved in that particular project.

PARAMETRIC MODELLING

BIM modelling tools are predominantly object-based, and the objects being modelled have a relationship within the model whereby any change in one object/element is automatically updated throughout the entire project. This is referred to as parametric modelling. The parametric attribute of BIM software should always be considered when making changes to elements within a model.

Some BIM applications such as REVIT Architecture, use 3 types of elements:

- Model elements: These are the elements that make up the 3D geometry of the building.
- Datum elements: Under this group are the *grids, levels,* and *reference planes* within the project.
- View-specific elements: These are elements which only appear in the view such as *tags, dimensions,* and *text annotations* (Autodesk, 2009).

MODEL SETUP

Accurate information about the BIM project needs to be sourced and used to set up the BIM model. However, sometimes it is possible to lack adequate drawing information to commence BIM modelling. This situation can affect the quality of the final BIM model especially if general approximations and assumptions are made. If proper CAD electronic drawings are not provided, effort should be made to acquire as much information as possible in order to set up the model. Under this circumstance, images of the floor plans can be to be imported into the BIM interface, scaled as accurately as possible and traced using the selected application.

BIM DATABASE

The BIM model consists of a database. This is where information about the project is stored and can be retrieved by stakeholders. The accuracy of the data will be based on architectural drawings and specifications which could differ from as-built drawings (if the building being modelled already exists), hence the need to post construction updates of the BIM model database.

OPEN STANDARDS DRIVEN BIM

Managing BIM in real-time is now possible using cloud computing on the Internet (Open Standards Driven BIM). Conditions of spaces and equipment can be shown in charts. When construction is complete the value of information from BIM becomes even more useful to maintain and monitor the lifecycle of the building.

5.4.1 Creating a new BIM model

When creating a model in a BIM environment, there are some important steps to follow. It is important to point out that some of these steps, as in 2D drafting, should be observed in a particular sequence while some can be applied at a later stage in the drawing process.

SPACE LAYOUT AND PROGRAM VALIDATION

The space layout in most BIM software programs is capable of providing multiple views of the work area. It is useful to begin by defining your work area and specifying the units of measurements, dimensions and detail level options. Also, project datums can be created, which might involve adding levels and providing labelled grid lines that can be adjusted to optimise the work area. The sheets/views will need to be organised if provisions are made for this in the BIM application being used.

The user can then proceed to model by establishing the site. This can be achieved by the importation of site contour data and its conversion to a 3D topographic representation. Identify The proposed position of the building within the site can then be identified and the footprint outlined, and the positions of adjacent site elements such as roads, fences, parking features, trees, etc. can be identified (Refer to the specific tutorial of the BIM application being used for detailed instructions regarding execution of the above.) See Figure 5.1 for an example of a BIM space layout.

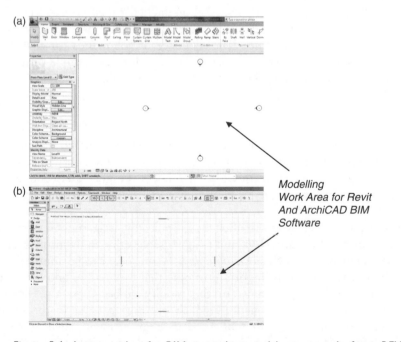

Figure 5.1 An example of a BIM space layout: (a) an example from REVIT Architecture; (b) an example from ArchiCAD.

MODEL CREATION

To add a wall, select the wall type and add to the work area; this updates on all viewports because most BIM applications have a parametric change engine, whereby changes made to any element in a particular view is updated throughout the entire project. It is therefore possible to build in code compliance because of the parametric attributes of some BIM applications.

To add doors to the floor plan, select the door tool and specify or select a predefined door type and place that door on the desired part of the wall. For some BIM software programs, such as REVIT, the door is attached automatically, creating the necessary opening and panels according to the specifications of the preselected door type. On placing the door, it is automatically numbered. Similar steps as those taken for the door can be applied to the windows. Roof modelling can be achieved by selecting a predesigned roof from the library and altering the dimensions as desired or by modelling the roof from scratch.

Some applications provide column elements already modelled. However, the columns can vary in type and be grouped according to families. In the majority of BIM applications, users can also create columns from scratch. (Refer to the specific tutorial of the BIM application being used for detailed instructions regarding execution of the above.)

Creation of roofs can vary depending on the type of roof. Roofs can be flat, pitched, domed, or any combination of these. For flat roofs, you create a slab and modify the slab construction to more accurately reflect the intended specification of the roof and slope character (which is gentle for flat roofs). For pitched roofs, you will need to draw ridge lines on the slab as required by the choice of pitched roof, select the angle of the slope and assign these properties. The sequence of commands for each task depends on the selected BIM application. See Figure 5.3 for an example of a BIM model where the commands listed in this section have been applied.

MODEL EDITING/MODIFICATION

In the majority of BIM applications, a separate 'modify' toolbar or ribbon is dedicated for this purpose. The parametric engine of BIM applications simplifies the model modification exercise; a change in any element within the model is immediately updated in the entire project (parametric attribute). Such modification tools are made up of:

- Properties side panel: Here you can alter the graphics by assigning values to the *view scale, graphic display, sun path,* and assign a scale value to the project. Also on this panel, identity data such as *view name, title on sheet* can be adjusted.
- Modify ribbon or toolbox: It is important to note that tool names and functions vary with different BIM programs. In the modify ribbon you also have a repetition of the modification functions which are also present in the menu. However, the tools include: *align tool,* used to align elements such as walls, windows, etc.; *move tool,* which is used to move elements; *offset tool,* which is used to offset lines; and *mirror tool,* which mirrors lines and objects, which can

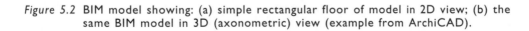

Figure 5.2 BIM model showing: (a) simple rectangular floor of model in 2D view; (b) the same BIM model in 3D (axonometric) view (example from ArchiCAD).

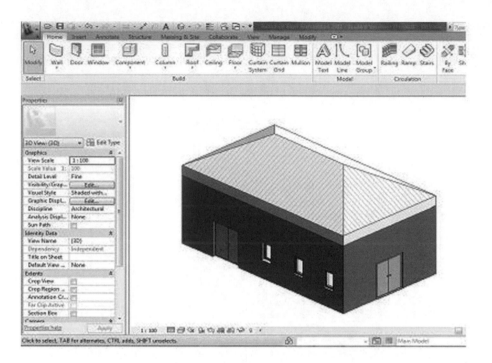

Figure 5.3 BIM model showing walls and roof in place with doors and windows inserted (example from REVIT Architecture).

help save time. Copy tool is also used to multiply existing lines and objects and also helps to save time, as existing lines and objects can be replicated rather than starting from scratch. The rotate tool is used to rotate an object about an axis to any desired angle. Examples can be seen in REVIT Architecture's modify ribbon (see Figure 5.4 (a)) and ArchiCAD's toolbox, which also has draw and annotation tools (see Figure 5.4 (b)).

The modification of elements such as walls can go beyond the creation of a simple wall to include the addition of BIM relevant wall details. Such a modification might include the addition of new layers of membrane to the wall and assigning attributes to them. The additional membranes can represent *insulation, structural elements, etc.*

SCHEDULE CREATION

Creating a door or window schedule: when it comes to creating a schedule, such as that for doors, windows, etc., the approach can vary for different BIM applications This process involves enriching the BIM database with information on the attributes and quantities of the elements/items in question, which is then manually or automatically organised by the software in a tabular form. This information can be used

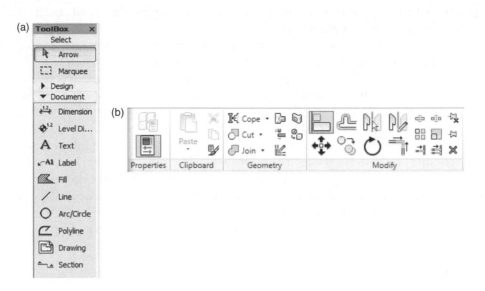

Figure 5.4 Active modelling tools can exist in software programs such as (a) the horizontal ribbon in REVIT Architecture or (b) the vertical toolbox in ArchiCAD.

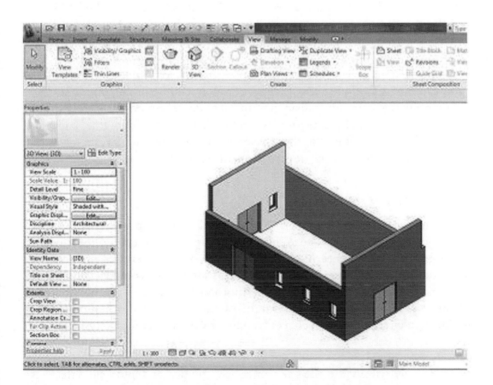

Figure 5.5 Example of a BIM model modified by raising the side walls using the modification tool (example from REVIT Architecture).

for quantity take-off (quantity of the material needed to complete the actual construction which can assist in cost estimates).

To create a door or window schedule: after creating the fields, you specify a sorting order, indicating the type of information you require. Such door or window information can include: width, height, operation character, frame type, frame material, fire rating and important comments made by the designers. The doors can be added to the project from a library and the values already assigned to the objects in the project automatically populate the door schedule.

Creating a room finish schedule: in order to create a room finish schedule you also will be required to create fields which in most cases will consist of name, base finish, wall finish, floor finish, ceiling finish, area, level and comments. You will then need to identify the rooms, such as the lounge, bedroom, etc. You will then be required to add the finish type from the library. It is important to mention that the method and sequence of executing this task can vary from one software program to another.

MODEL VISUALISATION

Model visualisation: Model visualisation is enhanced through rendering, which in turn is enhanced through the correct application of texture and lighting to the project. The application of materials and textures in BIM models is very similar to that of 3D. The difference, if any, might simply lie in the approach provided by the BIM software. These differences in rendering approach can also exist among 3D software programs.

After the selected materials have been applied to the different elements of the model, you can now modify the texture appropriately. While modifying your materials, it is important to maintain realism in terms of scale, reflection, transparency, etc. Materials can be assigned by category, whereby all objects within that particular category will reflect the attributes of the selected material. Alternatively, materials can be assigned by simply selecting the desired element within the project.

MODEL VALIDATION AND ANALYSIS

When a BIM model is complete, it can be used by all industry professionals participating in that particular project. The architects provide the architectural information, the engineer provides the engineering information, the building services engineer does the same, etc. (interoperability), and this makes the model quite complex. Because the BIM model is a complex repository of 3D information, there is the possibility of design elements either colliding with one another or cutting across each another when they shouldn't. In circumstances such as this, model validation and analysis becomes necessary.

There are some software packages which are designed to analyse architectural and engineering designs within the BIM for integrity, quality and physical safety. An example of such software is Solibri Model Checker. These 'model checking' programs are capable of analysing the BIM model and reveal potential flaws and weaknesses in the design, highlighting the clashing components and checking that the model complies with BIM requirements and the organisation's best practices.

It is important that the quality of the model is good because a poor quality model will result in poor quality analysis.

CODE COMPLIANCE CHECKS

Within the building industry of most developed countries there are enforced building codes which must be complied with. Code-compliance is a vital part of the design and it is possible to spend a tremendous amount of time reviewing plans for code-compliance. As BIM becomes more commonly used by building design teams, the relevance of the model for code analysis will become more important (see Section 5.2.1.2).

Discussion

- What is the difference between 3D CAD and BIM?
- What is the level of BIM uptake in your profession?
- Do you think the AEC/ FM industry should adopt BIM? If so, what are its barriers and how can these be overcome?

5.5 Summary

Although Building Information Modelling (BIM) has its roots in the mid-1980s, only recently has it risen in popularity within the architectural, engineering and construction facilities management (AEC/FM) industries.

With growing acceptance among the AEC/FM industries, BIM users are now able to represent their designs in a single 3D database which can be continually updated as changes are made or errors are detected during the design delivery stages and even up to the facilities management stage of the building in question.

BIM is used to produce a digital prototype of a building to understand its behaviour before construction. As a result, BIM aids communication between the design and construction teams. It is widely said that the advantages of BIM far outweigh its disadvantages, and, because BIM is fairly new, it will take time to realise its full potential. However, BIM adoption also has some disadvantages.

BIM models are similar to 3D models visually. However, the BIM model has more information and databases, is more accurate and requires more time to execute the model accurately. The adoption of BIM globally has been on the increase and software vendors are playing 'catch-up' to meet up with its growing popularity and demand.

Chapter 6

4D CAD

6.1 An introduction to 4D CAD

DEFINITION

2D, 3D CAD and Building Information Modelling (BIM) have been discussed in the previous chapters. This chapter concentrates at 4D CAD, which is sometimes referred to as 4D modelling. 4D CAD is a 3D model, or more specifically a 3D BIM model linked to the time dimension via a construction activity schedule. It enables the visualisation of the progression of the construction activities.

In a construction project, a 4D CAD simulates the process of transforming space over time and reflects the four-dimensional nature of construction activities. The development of 4D CAD involves the linking of a 3D graphic model or BIM model to a construction schedule. This integration process takes a 4D CAD model, which represents the product model (the design of a building/facility), and graphically incorporates the information traditionally represented in the construction schedule.

Unlike traditional construction planning tools, such as bar charts and network diagrams (which are generally static and show a number of events against a timeline), 4D CAD can represent and communicate the spatial and temporal aspects of construction schedules effectively, which allows project managers to create schedule alternatives to optimise a particular design. 4D CAD models that integrate physical 3D elements with time have been used to visualise construction processes in projects worldwide. It has demonstrated the benefits for the entire lifecycle of a project such as collaboration with stakeholders, making design decisions, assessing project constructability, identifying spatial conflicts in construction and so on.

However, there have been technical challenges in implementing the 4D CAD concept, such as recognition of the building elements in traditional 3D models, and lack of an automated linking process between building elements and construction schedule. With the success of BIM development in recent years, these challenges have been gradually removed by the advancement of BIM technology.

6.1.1 The historical development of 4D CAD

In 1910, the Gantt chart was developed by Henry Gantt, an American mechanical engineer, to monitor and record the progress of a project. It has been one of the key techniques for project planning for nearly 100 years. However, the Gantt chart, together with other techniques such as network diagrams and flow charts, does not communicate and visualise the project timeline. It is difficult for project managers to create alternative plans rapidly and to identify the best way to deliver a particular design. 4D CAD provides an answer. 4D CAD links to a construction activity schedule and enables a project manager to visualise the progression of the construction over time.

During 1990s, the 4D CAD model was introduced as a new presentation method (Cleveland, 1989). McKinney and Fischer (1998) developed the 4D CAD model as a planning tool to be used by a project team to investigate the impact of time and space during construction. Koo and Fischer (2000) highlighted 4D CAD's ability to visualise progress of the construction phase, by linking units of work to the work tasks on the construction schedule. Dawood et al. (2002) developed an integrated database to store connection information between construction activities and 3D objects using Standard Classification Methods (Uniclass). Wang et al. (2004) employed a work breakdown structure template to link with 3D CAD objects. Tanyer and Aouad (2005) extended the 4D model to the nD model to include project cost information.

Further applications demonstrated by Coles and Reinschmidt (1994) who created a 3D model over time to assist the planning process. Webb (2000) envisaged that the use of 4D simulations could assist in halving the waste costs associated with a construction project, and Chau et al. (2005) applied it in construction and resources management. This technology also has the potential for presenting ideas to clients in order to promote collaborative working (Fischer, 2001; Kähkönen and Leinonen, 2001), and to assist in the problems associated with site logistics and site layout (Zhang et al., 2000; Ma et al., 2005). Moreover, it can be used to improve site logistics, such as work execution space (Akinci et al., 2000; Heesom and Mahdjoubi, 2002) and to analyse the construction schedule to assess its executability (Koo and Fischer, 2000). Jongeling and Olofsson (2007) proposed a location-based schedule method which could be enhanced with the 4D CAD model to improve the work-flow onsite. Chin et al. (2008) used 4D CAD and RFID to monitor the progress of steelwork in a high-rise building. Park et al. (2011) investigated 4D CAD applicability for facility management.

Furthermore, the continuous development in technology and the emergence of Building Information Modelling (BIM), which was introduced in Chapter 5, have significantly advanced the development of 4D CAD. With the increasing adoption of BIM in architecture design, the use of 4D CAD is becoming more widely encouraged.

Discussion

- What are the key differences between 3D CAD and 4D CAD?
- What are the main drivers that are needed to happen in order for 4D CAD to be widely implemented?

6.2 4D CAD in practice

4D CAD technologies can be used by architects, engineers and contractors to analyse and visualise many aspects of a construction project, from the 3D design of a project and the sequence of construction to the relationships between schedule, cost and resources. Architects/engineers can create, update and maintain a 4D CAD model throughout a design/construction project. Different views of this 4D CAD model can be generated to clearly communicate the spatial and temporal aspects of construction schedules to all project participants. 4D CAD technologies will not just be used to simply animate the sequence of construction but will be used to communicate a wide range of project data much more clearly and efficiently. Architects, engineers and contractors can use 4D CAD environments to visually relate data much like the way engineers use gradated colour 3D models to visualise the stresses on structures. 4D CAD technologies could provide a view of an integrated project database: the database will store and maintain the representation of building components and construction activities, their interrelationships and relationships to other project data, such as cost, materials, etc. It will be designed to support concurrent engineering of the facility and its delivery process by supporting multiple representations of a single set of project data so that multiple applications from different disciplines can access and interpret the data. The database will not only support the use of 4D CAD technologies throughout the project lifecycle but also support other design and construction technologies.

6.2.1 4D CAD models in the project lifecycle

4D CAD models can be used throughout the project lifecycle. Typical projects can benefit from using 4D CAD models in the following design and construction stages (GSA, 2010):

- Preparation: 4D CAD models have been used for strategic project planning during the feasibility stage. For example, the model can be used to determine different phasing sequences and site configurations or to optimise the construction schedule. These models allow comparison of different alternatives with detailed assessment at a relatively low cost to the team and the client.
- Design and pre-construction: 4D CAD models can be used to improve the buildability of the design and to determine the advantages of different construction processes. These models can be used to optimise the phases in the construction

schedule. In addition, the models can be used to communicate the phasing plan to the client.

- Construction: 4D CAD models can also be used for the temporal aspects of construction, coordination and buildability. These include understanding where and how a subcontractor will work over a period of time and understanding traffic and site-flow processes. Onsite, these models can be used for regular construction progress reviews and to compare as-built with as-planned schedules for management and valuation purposes. 4D CAD models can also be used to communicate with clients, to enable them to review progress and visualise design changes during the construction process. In addition, 4D CAD models can be used to communicate the utility and control system changes required during specific periods and their impacts, especially for renovation projects.

6.3 The advantages of 4D CAD

4D CAD is capable of attaching time information to the traditional static 3D model, thus allowing planners/designers/developers to view construction progress or schedule in a 4D environment. 4D visualisation tools can demonstrate the entire construction progress in a vivid way and show potential conflicts in a construction site. What-if analysis can also be exercised to assess and compare several planning options in order to select a better strategy.

In summary, the advantages of using 4D CAD modelling may include:

- In design and pre-construction:
  - Increased stakeholder communication through visualisation and better understanding of the design proposal.
  - Preliminary analysis of site layout, construction phasing, construction activities and site configuration.
  - Supporting the development of construction sequencing optimisation.
  - Better integration and cost estimate/budget control.
  - Early detection of design conflicts and possible resolutions.
- In construction:
  - Improved subcontractor coordination.
  - Reduced number of design changes during construction.
  - Detailed analysis of construction sequencing for potential conflicts.
  - 4D CAD models can help with enabling stakeholders to visualise the construction process and to monitor the construction progress.

Moreover, 4D CAD can improve safety management on site: a 4D CAD model can be analysed together with safety rules to automatically detect any working-at-height hazards and also to indicate any necessary safety measures in terms of activities and requirements (Benjaoran and Bhokha, 2010). These safety measures can be included in the construction schedule and visualised on the 4D CAD together with the other construction sequences.

6.4 The limitations of 4D CAD

Although 4D CAD modelling has demonstrated clear advantages during the design and construction process, a number of limitations have been highlighted during the practical implementation of the technology (Benjaoran and Bhokha, 2009).

- A 4D CAD model requires integration between design and planning information. Users often have to switch between two or more software applications to acquire all the information they need to produce a 4D model. However, the latest developments in specialised software is making the process more efficient.
- A 4D CAD model can provide a virtual experience of the work to the team. However, the visualisation of the 4D CAD model is still unable to present all the information of a construction plan as not all the construction activities can be visually presented. Also, unlike traditional graphical methods of monitoring and recording the project, such as a Gantt chart, users can see the whole programme at the same time. When the project is large and complex, the overview of the whole project is important to the project team.
- The duration of the activity is not directly presented in a quantitative manner in a 4D graphical model. It is often difficult to evaluate and compare the duration of activities, particularly when the project is large or the activities have very different start times. On the other hand, the Gantt chart can illustrate the activity duration via both the length of a bar and a number. The activity duration is important information for the schedule. It helps users justify the reasonableness of the schedule and the difficulty of an activity. Most recent 4D modelling tools have started to include the Gantt chart view in their system.
- The 4D CAD model cannot easily present the activities' relationships (such as a predecessor, a successor, a predecessor and a successor), and the Gantt chart view is still required.
- Developing the models can be very time consuming and errors cannot be easily identified by the software. Human errors can be made and difficult to identify when linking the 3D model and the schedule, particularly in large and complex project.
- 4D CAD models cannot modify or optimise the schedule automatically, and manual interaction with the project team is still required to fully realise its benefits.

6.5 The 4D CAD modelling process

4D modelling process can be broken down into the following steps:

1   Prepare a 3D CAD model from the 2D CAD drawings or BIM models (integrate architecture, structure and MEP information);
2   Prepare the construction schedule, which includes all project activities in the project;
3   Link 3D CAD/BIM objects with construction activities through linking keys (e.g. activity name, layer, object name, etc.);
4   Update and maintain the 4D CAD model.

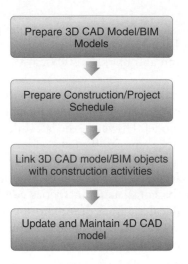

Figure 6.1 The 4D CAD modelling process.

6.5.1 The CAD modelling process

- Prepare a 3D CAD model/BIM model
  3D CAD/BIM models have a specific layering scheme within the 3D modelling application. In the actual construction of the project, building elements are never constructed all at once. 3D models are not usually specially designed for 4D scheduling, therefore it is essential for the model management to have a 3D model layering scheme which supports the 4D CAD modelling activities. This may involve:

  - Organising the 3D model into work elements to match the level of detail in the schedule
  - Placing the geometric information into different layers
  - Regrouping the CAD components onto different set of layers
  - Breaking component into some small pieces onto different layers.

  When the building elements in the 3D model are not sufficient to describe the construction process, supporting building elements and activities, such as scaffolding, need to be created. BIM models have largely resolved the problems associated with converting 3D objects to a 4D model as the object hierarchy in a BIM model is more clear and relevant to a construction schedule. Nevertheless, certain reorganisation of the BIM objects might be still required.
- Prepare construction/project schedule
  Construction schedules nowadays are prepared by the project management team with a scheduling tool such as Primavera or Microsoft Project. Typical tasks for generating the construction schedule include:

- – Generating and naming the project activities
- – Assigning the duration of the activities
- – Assigning the relationship between activities
- – Sequencing the activities.

For the purpose of 4D CAD modelling, it is worth noting that the schedule should be at the same level of detail as the 4D visualisation. It is advisable to categorise the actions into different types, because not all activities can be shown in the 4D visualisation.

- • Linking the 3D model and the schedule

  If the 3D model and schedule are set up correctly, the linking of the two components together to create the 4D CAD model should be a relative straightforward process. However, if the project is large and complex, it is still going to be a time-consuming task, because a manual link between the building components and the schedule is still required in most cases. The linking process can be done based on the component name, layer or even construction method. Recently, some software applications have been developed to offer an automated linking process based on the unique identifiers defined in both 3D modelling and 4D modelling software.
- • Update and maintain the 4D CAD model

  Throughout the construction project, the 3D design model and construction schedule can be revised and the 4D CAD model needs to be updated accordingly. This could include: regrouping the 3D components, changing an activity's name and duration, resequencing activities, exporting and importing the revised 3D model and relinking the 3D object with the activities. It can be very cumbersome and difficult for the project team.

6.5.2 4D modelling software

Over the years, many software applications have been developed to support 4D modelling; they typically come within a suite of applications or as a stand-alone third-party application. 4D CAD modelling applications that are within a suite of applications allow the project team to create the 3D and 4D model all within one application family without having issues of data interoperability. 3D components are linked with time by either specifying specific phases within the modelling application, or by importing a project schedule into the application. Stand-alone 4D modelling applications will import both the 3D model and project schedule and the linkages are then created in these applications.

In the following section, we will introduce several 4D CAD software applications, which are currently available on the market.

- • ProjectWise Schedule Simulation – part of Bentley Navigator: ProjectWise Schedule Simulation provides further insight into critical project planning information by importing, and linking to, schedule information managed in Microsoft Project, Excel or Primavera. With ProjectWise Schedule Simulation you can visually explore alternatives and create the most cost-effective and safest construction scenarios by visualising schedule information and animating your 3D engineering model based on project schedule data (Bentley, 2010).

- Naviswork Timeliner – part of Autodesk Naviswork Suite: The TimeLiner tool adds 4D schedule simulation to Autodesk Navisworks Manage. TimeLiner imports schedules from a variety of sources, allows you to connect objects in the model with tasks in the schedule and simulate the schedule showing the effects on the model, including planned against actual schedules. TimeLiner also allows the export of images and animations based on the results of the simulation. TimeLiner will automatically update the simulation if the model or schedule changes (Beyond the Paper, 2010).

- Innovaya Visual Simulation integrates Building Information Models (BIM) objects with scheduling activities to perform 4D construction planning and constructability analysis. It effectively improves project communication, coordination and construction logistics planning. With its robust 3D engine and extremely easy-to-use interface, Visual 4D Simulation helps you build optimised task sequences to achieve project time savings (Innovaya, 2010).

6.5.3 4D modelling, a practical example

A 4D CAD modelling example is provided as a demonstration of how to create a simple 4D CAD model using the following software (although proprietary software is used here, the activities of creating a 4D CAD model can be applied to other applications):

- 3D model preparation – Autodesk Revit Architecture 2011
- Project Scheduling – Microsoft Project 2007
- 4D modelling – Autodesk Naviswork Manage 2012.

The modelling process

- Develop a 3D model using Autodesk Revit and export the file to a Navisworks file. The Naviswork exporter for Revit is freely available from the Autodesk web site. The Naviswork exporter is located under 'Add-in' in the external tools menu (see Figure 6.2). The 3D view of the model has to be selected to enable the export function.
- Develop a CPM schedule for the project in MS Project 2007 (Figure 6.3).
- Open Navisworks and then open the model exported from Revit. Revit exported the file as a Navisworks cache file (.nwc file) (see Figure 6.4).
- Open the Navisworks Timeliner module by selecting Timeliner menu item. (see Figure 6.5).
- Go to the 'Data Source' tab and then add the Microsoft Project file (Figure 6.6). You will then need to right click the linked file and select 'Rebuild task hierarchy' so that the activities in the scheduler are imported (see Figure 6.7).

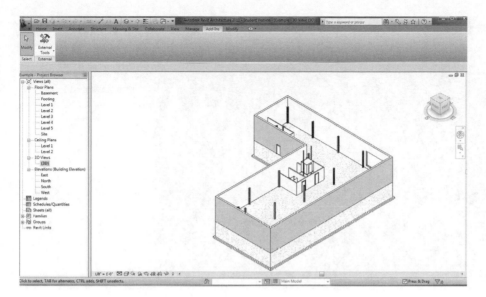

Figure 6.2 Model showing 3D view in Revit with external tools export function.

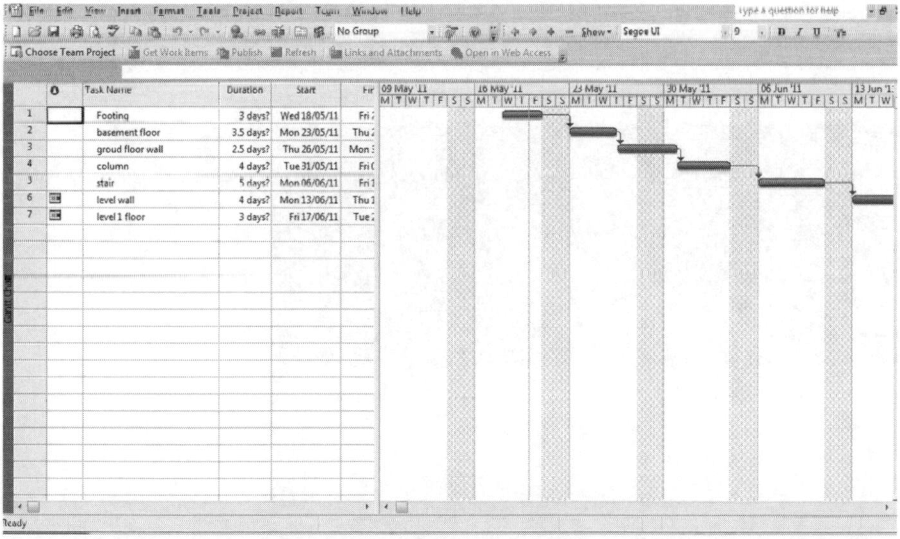

Figure 6.3 Construction schedule in Microsoft Project.

Figure 6.4 Model opened in Naviswork

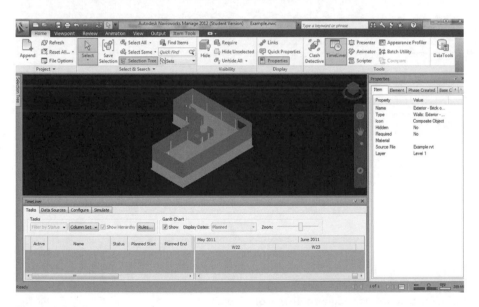

Figure 6.5 Naviswork Timeliner options

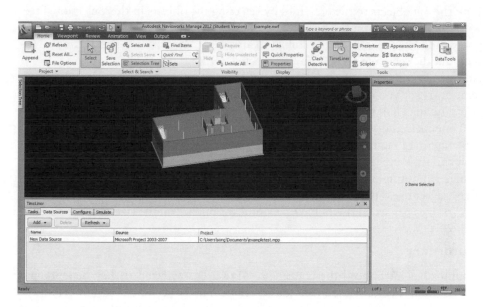

Figure 6.6 Naviswork Timeliner import schedule from Microsoft Project.

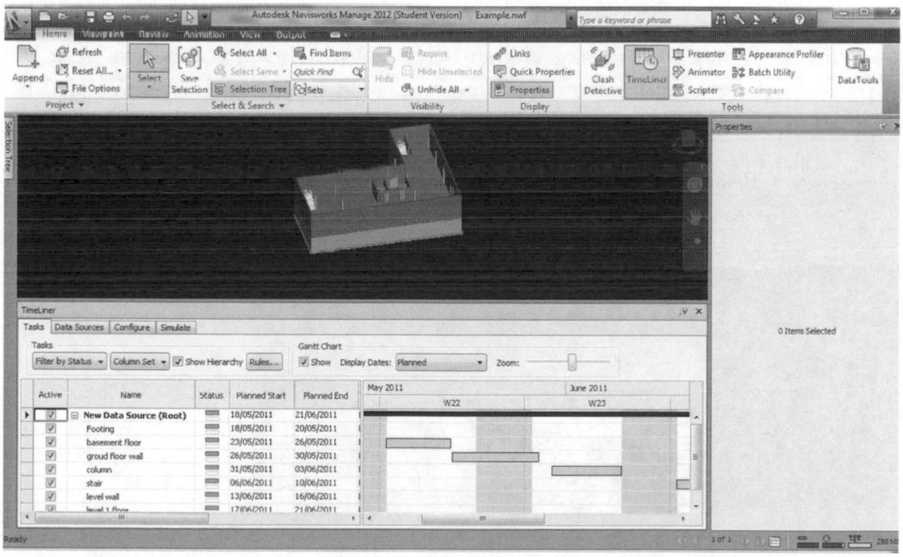

Figure 6.7 Construction schedule opened in Naviswork Timeliner.

- Link the 3D objects to the activities by selecting the objects in the 'Selection Tree' window or directly on the 3D objects and then right click on the activity which you would like the object to be linked to and select 'Attach current selection'. This will link the schedule activity to the selected 3D objects.
- Add an activity type to each of the activities, e.g. construction, demolition or temporary, or you can create your own activity types (see Figure 6.8).
- After linking all of the objects, you can review the simulation in the 'Simulate' tab of Timeliner, and you can revise the playback settings in the 'Configure' tab of Timeliner (see Figure 6.9).
- You can then export the 4D model as a video file. To do this, go to 'File – Export – Image and Animation' and select 'Animation', and within the source dropdown menu select the 'Timeliner Simulation'. Note that the size of the window and the FPS (frames per second) will have a significant impact on the size of the video file.

6.6 Summary

This chapter has introduced the concept of 4D CAD, reviewed its key benefits and limitations, and demonstrated the basic modelling process. It is important to understand that this chapter only provides a starting point to the full appreciation of 4D CAD technology.

There are serveral challenges to be overcome in the development of 4D CAD, such as the recognition of 3D objects in relation to building elements, the automatic linking of construction activities to building elements, etc. With the introduction of BIM, many of these challenges have been overcome, and BIM-based CAD systems will be primarliy used in 4D modelling in the future.

Figure 6.8 Naviswork Timeliner set activity type.

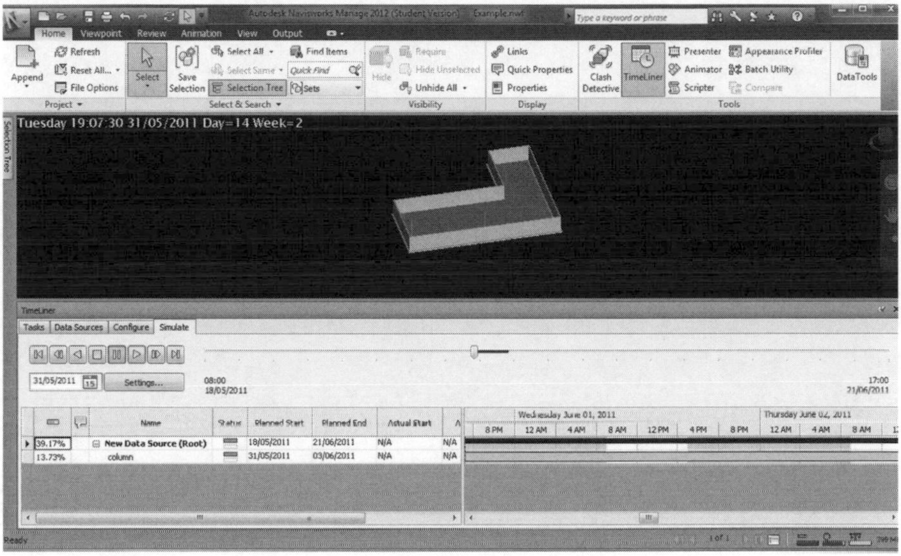

Figure 6.9 Naviswork Timeliner simulation.

Chapter 7

nD modelling

7.1 Introduction

So far, 2D, 3D and 4D CAD modelling have been introduced in the previous chapters. A rhetorical question now remains: what lies beyond 4D modelling? This was touched upon in Chapter 2, where it was noted that the merging of process and production modelling led to nD modelling. The concept of nD modelling was first introduced by the authors at the University of Salford. In this chapter, we will focus on the concept of nD modelling, its development and its future.

One of the main challenges facing the construction industry today is how to improve the efficiency and effectiveness of the integrated design and construction process. Moreover, what contribution can the effective use of information technology make to this?

Designing a building is part art and part alchemy. It is no longer simply a question of organising a range of facilities on a single site; the needs of a whole host of project stakeholders have to be satisfied. Thus, the way in which the building will fulfil the expectations of all those parties who form the spectrum of building stakeholders is increasingly becoming the measure of its success. The stakeholders include not only the organisations and individuals who occupy the building, but also those who have provided it, those who manage it and those who live with it – the community in general.

The design not only has to be buildable (in terms of cost and time), but stakeholders are increasingly enquiring about its maintainability, sustainability, accessibility, crime deterrent features, and its acoustic and energy performances. Often, a whole host of construction specialists are involved in instigating these aspects of design, such as accessibility auditors, FM specialists and acoustic consultants. With so much information, and from so many experts, it becomes very difficult for the client to visualise the design, any changes that are applied, and the subsequent impact on the time and cost of the construction project. Changing and adapting the design, planning schedules and cost estimates to aid client decision-making can be laborious, time-consuming and costly. Each of these design parameters that the stakeholders seek to consider will have a host of social, economic and legislative constraints that may be in conflict with one another.

Furthermore, as each of these factors vary – in the amount and type of demands they make – they will have a direct impact on the time and cost of the construction project. The criteria for successful design therefore will include a measure of the extent to which all these factors can be co-ordinated and mutually satisfied to meet the expectations of all the parties involved.

Traditionally, specialist input of each of these design criteria is usually undertaken in a sequential step-by-step fashion, whereby the design proceeds following a number of changes after satisfying the legal requirements of a specialist consultant, and then continues on to the next consultant who, in turn, makes a number of design recommendation changes. In this sense, specialist design changes are made in isolation from each other, creating an over-the-brick-wall-effect, where each discrete change by one consultant plays little or no regard to the next (see Figure 7.1). Therefore, it is often difficult to balance the design between aesthetics, ecology and economy – a three-dimensional view of design that acknowledges its social, environmental and economic roles – in order to satisfy the needs of all the stakeholders.

These problems are mainly attributed to the vast amount of information and knowledge that is required to bring about good design and construction co-ordination and communication between a traditionally fragmented supply chain. The complexity of the problem increases with the fact that this information is produced by a number of construction professionals of different backgrounds, as discussed in Chapter 2. Therefore, without an effective implementation of IT and processes to control and manage this information, the problem will only expound as construction projects become more and more complex, and as stakeholders increasingly enquire about the performance of buildings (sustainability, accessibility, acoustic, energy, maintainability, crime, etc.).

Current developments that aim to address this problem have been led by two factors, namely a technology push and a strategic pull. The technological push has led to the development and implementation of hardware and software to improve a number of functions, including the development of building information models (integration of 3D + time/4D modelling, for example), and information standardisation. The strategic pull, on the other hand, can be gauged by the growing number of workshops, seminars and conferences on the subject.

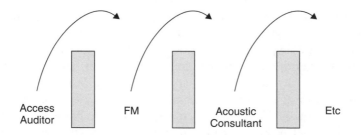

Figure 7.1 Sequential over the 'brick wall' approach.

7.2 What is nD modelling?

An nD model is an extension of the building information model that incorporates all the design information required at each stage of the lifecycle of a building facility (Lee et al., 2003). Thus, a building information model (BIM) is a computer model database of building design information, which may also contain information about the building's construction, management, operation and maintenance (Graphisoft, 2003). From this database, different views of the information can be generated automatically, views that correspond to traditional design documents such as plans, sections, elevations and schedules. As the documents are derived from the same database, they are all coordinated and accurate – any design changes made in the model will automatically be reflected in the resulting drawings, ensuring a complete and consistent set of documentation (Graphisoft, 2003). It builds upon the concept of 2D, 3D and 4D.

nD modelling develops the concept of 4D modelling and aims to integrate an nth number of design dimensions into a holistic model, which would enable users to portray and visually project the building design over its complete lifecycle. nD modelling is based upon the building information model (BIM), a concept first introduced in the 1970s and the basis of considerable research in construction IT ever since. The idea evolved with the introduction of object-oriented CAD; the 'objects' in these CAD systems (e.g. doors, walls, windows, roofs) can also store non-graphical data about the building in a logical structure. The BIM is a repository that stores all the data 'objects' with each object being described only once. Both graphical and non-graphical documents, such as drawings and specifications, schedules and other data respectively, are included. Changes to each item are made in only one place and so each project participant sees the same information in the repository. By handling project documentation in this way, communication problems that slow down projects and increase costs can be greatly reduced (Cyon Research, 2003).

Leading CAD vendors such as AutoDesk, Bentley and Graphisoft have promoted BIM heavily with their own BIM solutions and demonstrated the benefits of the concept. However, as these solutions are based on different, non-compatible standards, an open and neutral data format is required to ensure data compatibility across the different applications. Industry Foundation Classes (IFC), developed by the International Alliance for Interoperability (IAI), provides such capabilities. IFC provides a set of rules and protocols that determines how the data representing the building in the model are defined and the agreed specification of classes of components enables the development of a common language for construction. IFC-based objects allow project models to be shared whilst allowing each profession to define its own view of the objects contained in that model. This leads to improved efficiency in cost estimating, building services design, construction, and facility management: IFC enable interoperability between the various AEC/FM software applications, allowing software developers to use IFC to create applications that use universal objects that are based on the IFC specification. Furthermore, this shared data can continue to evolve after the design phase and throughout the construction and occupation of the building. IFC has been widely accepted by all major software vendors, and most of the BIM solutions nowadays support IFC data exchange.

7.3 nD modelling research development

nD modelling research at the University of Salford has developed a multi-dimensional computer model that will portray and visually project the entire design and construction process, enabling users to 'see' and simulate the whole-life of the project. This, it is anticipated, will help to improve the decision-making process and construction performance by enabling true 'what-if' analysis to be performed in order to demonstrate the real cost in terms of the variables of the design issues (see Figure 7.2). Therefore, the trade-offs between the parameters can be clearly envisaged:

- Predict and plan the construction process (4D CAD modelling)
- Determine cost options
- Maximise sustainability
- Investigate energy requirements
- Examine people's accessibility
- Determine maintenance needs
- Incorporate crime deterrent features
- Examine the building's acoustics.

The aim of the research was to develop the infrastructure, methodologies and technologies that would facilitate the integration of time, cost, accessibility, sustainability, maintainability, acoustics, crime, and thermal requirements. It assembled and combined the leading advances that had been made in discrete information communication technologies (ICTs) and process improvement to produce an integrated prototyping platform for the construction and engineering industries. This output will allow seamless communication, simulation and visualisation, and intelligent and dynamic interaction of emerging building design prototypes, so that their fitness of purpose for economic, environmental, building performance, human usability will be considered in an integrated manner. Conceptually, this will involve taking three-dimensional modelling in the built environment to an nth number of dimensions.

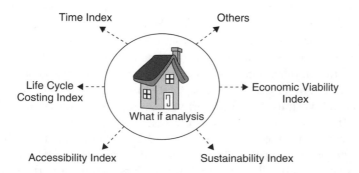

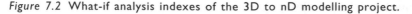

Figure 7.2 What-if analysis indexes of the 3D to nD modelling project.

The developed nD tool builds upon the concept of BIM and is IFC-based: the system architecture is illustrated in Figure 7.3:

- nD knowledge base: a platform that provides information analysis services for various design perspectives of the nD modelling (i.e. accessibility requirements, crime deterrent measures, sustainability requirements, etc.). Information from various design handbooks and guidelines on the legislative specifications of building components will be used together with physical building data from building information models to perform individual analysis (see Figure 1.4).
- Decision support: multi-criterion decision analysis (MCDA) techniques have been adopted for the combined assessment of qualitative criteria (i.e. criteria from the Building Regulations and British Standard documents that cannot be directly measured against in their present form) and quantitative criteria (e.g. expressed in geometric dimensions, monetary units, etc.). Analytic hierarchy process (AHP) is used to assess both qualitative criteria (i.e. criteria that cannot be directly measured) and quantitative criteria (e.g. expressed in dimensions, monetary units, etc.). The accessibility assessment model based on the AHP methodology is developed to support the decision making on accessible design (see Figure 1.5).

7.4 The future of nD modelling

The 3D to nD modelling project defined the need, vision and technology to bring about an improved decision-making/what-if analysis prototype tool. It harmonised the constraints and conflicts acting upon construction stakeholders, which are imposed by

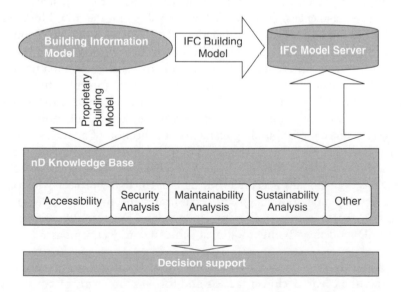

Figure 7.3 System architecture of the nD modéling prototype tool.

a number and a variety of social, economic and legislative factors, different episte-mological backgrounds and methods, and the personal requirements and agendas of other interested stakeholder groups. This was achieved by creating a tool that assesses different design criteria (namely whole-lifecycle costing, acoustics, environmental impact, crime and accessibility) required at each stage of the lifecycle of a building facility. IFC was utilised to enable cross-platform sharing of design information. The benefits of the prototype were demonstrated using the new extension of The Lowry, a national landmark millennium building designed by Michael Wilford & Partners to house the work of the artist L.S. Lowry. Since the onset of the 3D to nD modelling project, the terms 'nD modelling' and 'nD CAD' have gained increased usage in the field of information communication technologies (ICTs) and has been a leading track session at six international conferences. The value and need of nD modelling was demonstrated by the 3D to nD modelling project. Research around the world has continued to develop 4D (3D + time, 3D + cost, 3D + accessibility, 3D + energy, etc.) and 5D models somewhat discretely (Heesom and Mahdjoubi, 2004; Fischer and Liston, 2001).

The AEC/FM industry faces considerable economic, environmental and societal challenges as the twenty-first century ensues. Technology research, development and deployment will be vital to meeting these challenges and seizing opportunities for future growth. Increased globalisation of markets, for example, will create new market opportunities, but will require the development of more advanced technology to ensure the industry is globally competitive. Systems/processes that are resource efficient, cost-effective, and environmentally-sound will be the cornerstone in maintaining this balance. The 3D to nD modelling research project, with the participation of over 150 leading international researchers from academe and industry in five workshops, reviewed and analysed the factors affecting the competitiveness of the industry and its ability to meet future challenges. The workshops resulted in the nD Modelling Roadmap: A Vision for nD-Enabled Construction (Lee et al., 2005), a visionary document that identifies the major needs and challenges over the next two decades. To meet the goals for the twenty-first century, the roadmap advocates focus on a number of key issues, such as scalability, different actors/users, etc., to truly mimic the 'traditional' design process. The future of IT in construction must imitate the way in which architects unconsciously think about buildings while they are designing: as a collection of spaces, a collection of building elements, a collection of systems, a series of layers, etc. (Lawson, 2004). When designing, architects oscillate without noticing between these descriptions of the building, adopting parallel lines of thought. Now that the concept of nD has been globally accepted (through the 3D to nD project), the focus needs to turn to developing a tool that fully orchestrates the process of design of modern buildings so that it can be easily implemented by practitioners. Thus, the second generation nD modelling 2G:nD tool must simultaneously encompass:

- Embrained knowledge: encompassing the viewpoints of different stakeholders/ users of nD (i.e. the client, architect, access auditor, etc.) in terms of both feed-forward and feedback of design information. Thus, it will be actor configurable.

- Process knowledge: so that nD can harness and be harnessed within various operating schematics, such as the business process, design and construction process, the operational/maintenance process, etc. Thus, it will be process configurable.
- Encoded knowledge: ensuring the design conforms to the respective design standards in accordance to the building type. Thus, it will be code configurable.
- Embodied knowledge: enabling the scalability of use, from a single building to city and urban environments. Thus, it will be scale configurable.
- Moreover, in doing so, the nD tool should be self-learning. The embrained, process, encoded and embodied knowledge cells will, in turn, be self-improving, more intuitive and pro-active. Thus, it will cultivate knowledge.

See Figure 7.4, which shows the future nD modeling research framework that was developed through five workshops (see Lee et al., 2005).

Therefore, the next phase of nD modelling is to build intelligence into the nD environment by using multi-agent technology so that it orchestrates the design process simultaneously or as required. The nD prototype tool so far encapsulates the encoded knowledge domain. Although state-of-the-art, it can be duly criticised as being somewhat tedious to use, as the process of information retrieval and analysis is manual. Therefore, with the paramount need for nD modelling in the twenty-first century, the primary objective of 2G:nD is to automate this process using intelligent agent technology in order to develop a holistic nD environment that incorporates all aspects of the nD modelling research framework (Figure 7.4). It is proposed that different

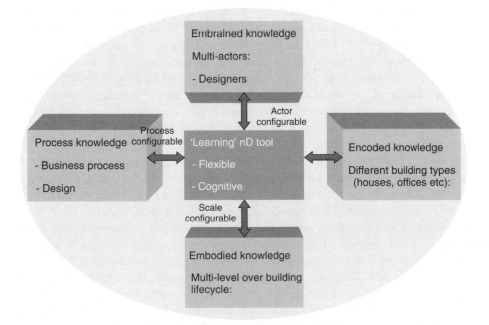

Figure 7.4 nD modelling research framework.

users/actors can utilise nD technology from their own discipline perspective (i.e. drawings for designers, cost spreadsheets for quantity surveyors, etc.); it can be code adapted for different building types (i.e. check conformity for housing, offices, etc.); it can be scaled accordingly to tackle buildings or urban environments; it will address various processes (i.e. business, design, facilities management, etc.); and will be self-learning by using a number of technologies and techniques (i.e. mobile computing, pervasive computing, etc.).

In the proposed 2G:nD modelling environment, a multi-agent system (MAS) will be developed to simulate the multi-disciplinary construction project environment, which will be intelligent and intuitive. MAS is based on the idea that different independent software components/agents can cope with problems which are hard to solve using the traditional centralised approach to computation, and whose collective skills can be applied in complex and real-time domains (Ferber, 1999). The target of such systems is to demonstrate how goal-directed, robust and optimal behaviour can arise from interactions between individual autonomous intelligent software agents. Over the last decade, a number of successful applications have been developed in the field of business processes (Jennings et al., 1996), in electricity management (Cockburn and Jennings, 1996; Jennings et al., 1996), and control (Atkins et al., 1999; Azevedo et al., 2000). Real-time industrial applications have also been built (Parunak, 1999; Gallimore et al., 1999) either by upgrading existing stand-alone applications with agent technology (Jennings et al., 1996) or by building complex, adaptive and dynamic multi-agent systems from first principles. To date, MAS has had very limited application in conflict resolution of the design of modern buildings – early research, such as ADLIB, investigated agent negotiation for collaborative design (Newnhan et al., 1999). 2G:nD aims to develop separate modules (i.e. actors, code, scale and process configurations that will 'learn' as they are being used), whereby each one will provide a discrete solution, and/or then be allowed to proactively cooperate, negotiate and exchange information autonomously in order to solve holistic what-if analysis problems within a time period and level of detail that is conducive to the user. This will overcome the manual analysis that is apparent within the existing 3D to nD modelling prototype.

Moreover, to specifically combat the embrained/multi-actor perspective, simulation using virtual characters (avatars) will be developed to provide potential users with a 'real' look and feel of the design in a virtual environment. Virtual characters have been an important part of computer graphics in recent years; characters have often taken forms such as synthetic humans, animals, mythological creatures, and non-organic objects that exhibit life-like properties (walking lamps etc.). They have advanced dramatically over the past decade, revolutionising the motion picture, game and multimedia industries. In the built and human environment, virtual characters have largely been used for crowd simulation in emergency egress and pedestrian simulation (Musse and Thalmann, 2001; Penn and Turner, 2002). In the 2G:nD project, we aim to bring intelligence to the virtual characters. The behaviour of the virtual characters can be modelled and defined. It is presently challenging to define all aspects of the behaviour of a complex virtual character; the desired behaviour may be impossible to define ahead of time if the characters of the virtual world change in unexpected or diverse ways (Dinerstein et al., 2004). For these reasons, it is desirable to make virtual characters as autonomous and intelligent as possible whilst still maintaining animator control over their high-level goals. This can be accomplished with a behavioural model: an executable model defining how the character should react to stimuli from its environment, and/or by a cognitive model: an executable model of the characters' thought processes (Tu and Terzopoulos, 1994). Both modelling techniques will be adopted to define the behaviours of potential users (e.g. children, elderly, disabled, etc.). Thus, the use of 2D CAD is still widespread in the UK AEC/FM industry, as described in Chapter 3, but we are still far from harnessing the advantages of BIM and the potential of nD. However, the future described here presents a desirable and much needed goal.

Bibliography

Adachi, Y. (2002) 'Overview of IFC model server framework'. In Proceedings of the International ECPPM Conference – eWork and eBusiness in Architecture, Engineering and Construction, Portoroz, Slovenia.

Akinci, B., Fischer, M., Levitt, R. and Carlson, R. (2000) 'Formalization and automation of time-space conflict analysis'. Working Paper No. 59, CIFE, Stanford.

Alexander, C. (1964) *Notes on the Synthesis of Form*. Cambridge, MA: Harvard University Press.

Alshawi, M. (1996) 'SPACE: integrated environment'. Internal Paper, University of Salford, July 1996.

Alshawi, M. and Ingirige, B. (2003) 'Web-enabled project management: an emerging paradigm in construction'. *Automation in Construction*, 12, 349–364.

Alshawi, M., Faraj, I., Aouad, G., Child, T. and Underwood, J. (1999) 'An IFC web-based collaborative construction computer environment'. Invited paper, in Proceedings of the International Construction IT Conference, Construction Industry Development Board, Malaysia, pp. 8–33.

Ammermann, E., Junge, R., Katranuschkov, P. and Scherer, R. J. (1994) *Concept of an Object-Oriented Product Model for Building Design*. Dresden: Technische Universität.

Anderson, E. J. (1994) *Management of Manufacturing, Models and Analysis*. Wokingham: Addison-Wesley.

Aouad, G. (1999) 'Trends in information visualisation in construction'. Proceedings of the 1999 International Conference on Information Visualisation IV 99, London.

Aouad, G., Betts, M., Brandon, P., Brown, F., Child, T., Cooper, G., Ford, S., Kirkham, J., Oxman, R., Sarshar, M. and Young, B. (1994) 'Integrated databases for design and construction: final report'. University of Salford (internal report), July 1994.

Aouad, G., Brandon, P., Brown, F., Child, T., Cooper, G., Ford, S., Kirkham, J., Oxman, R. and Young, B. (1995) 'The conceptual modelling of construction management information'. *Automation in Construction*, 3, 267–282.

Aouad, G., Hinks, J., Cooper, R., Sheath, D., Kagioglou, M. and Sexton, M. (1998) 'An Information Technology (IT) map for a generic design and construction process protocol'. *Journal of Construction Procurement*, November, 4(1), 132–151.

Aouad, G., Kagioglou, M. and Cooper, R. (1999) 'IT in construction: a driver or and enabler?' *Journal of Logistics and Information Management*, 12, 130–137.

Aouad, G., Kirkham, J., Brandon, P., Brown, F., Cooper, G., Ford, S., Oxman, R., Sarshar, M. and Young, B. (1993) 'Information modelling in the construction industry – the information engineering approach'. *Construction Management and Economics*, 11(5), 384–397.

Aouad, G., Sun, M. and Faraj, I. (2002) 'Automatic generation of data representations for construction application'. *Construction Innovation*, 2, 151–165.

Aranda-Mena, G., Crawford, J., and Chavez, A., and Froese, T. (2008) 'Building information modelling demystified: does it make business sense to adopt BIM?' CIB-W78 25th International Conference on Information Technology in Construction, 2 (3), 419–433.

Arditi, D. and Gunaydin, H. M. (1998) 'Factors that affect process quality in the life cycle of building projects'. *Journal of Construction Engineering and Management*, 124(3), 194–203.

Atkins E. M., Durfee E. H. and Shin, K. G. (1999) 'Autonomous Flight with CIRCA-II. In Autonomous Agents '99 Workshop on Autonomy Control Software, May.

ATLAS (1992) Architecture, Methodology and Tools for Computer Integrated Large Scale Engineering – ESPRIT Project 7280, Technical Annex Part 1, General Project Overview.

Attar, R., Hailemariam, E., Glueck, M., Tessier, J., and Khan, A. (2010) BIM-based Building Performance Monitor. Invited Video: SimAUD 2010-Orlando, FL, USA. [Online] Available at: <http://www.autodeskresearch.com/pdf/09_cBIMBasedBuildingDashboard.pdf> [Accessed February 2010].

Augenbroe, G. (1993) COMBINE, Final Report. Delft University.

Autodesk. (2009) Revit Architecture Tutorial. Autodesk. [Online] Available at: <http://usa.autodesk.com/adsk/servlet/item?siteID=123112&id=11091739> [Accessed January 2011]

Azevedo, C. P., Feiju, B. and Costa, M. (2000) 'Control centers evolve with agent technology'. *IEEE Computer Applications in Power*, 13(3), 48–53.

Banwell, H. (1964) *Report of the Committee on the Placing and Management of Contracts for Building and Civil Engineering Work*. London: HMSO.

Bedrick, J. (2005) 'FireGL BIM and process improvement'. *Aecbytes*. [Online] Available at: <http://www.aecbytes.com/viewpoint/2005/issue_20.html> [Accessed February 2011].

Benjaoran, V. and Bhokha, S. (2009) 'Enhancing visualization of 4D CAD model compared to conventional methods'. *Engineering, Construction and Architectural Management*, 16(4), 392–408. [Online] Available at: http://www.emeraldinsight.com/10.1108/0969998091097 0860 [Accessed 20 May 2011].

Benjaoran, V. and Bhokha, S. (2010) 'An integrated safety management with construction management using 4D CAD model'. *Safety Science*, 48(3), 395–403.

Bentley (2010) 'Dynamic project review and analysis'. [Online] Available at: http://www.bentley.com/en-US/Products/ProjectWise+Navigator/Schedule+Simulation.htm [Accessed 1 September 2010].

Beyond the Paper (2010) 'Autodesk Navisworks Manage 2010 – new features'. [Online] Available at: http://dwf.blogs.com/beyond_the_paper/2009/04/new-features-of-autodesk-navisworks-manage-2010.html [Accessed 1 September 2010].

Bjork, B. C. (1989) 'Basic structure of a proposed building product model'. *Computer Aided Design*, 21(2), 71–78.

Bjork, B. C. (1991) 'A unified approach for modelling construction information'. *Building and Environment*, 27(2), 173–194.

Björk, B. and Laakso, M. (2010) 'CAD standardization in construction – a process view'. Special issue on BIM standardization'. *Automation in Construction*, 19(4), 398–406.

Bjork, B. C. and Wix, J. (1991) *An Introduction to STEP*. Bracknell, England: VTT and Wix McLelland Ltd.

Bofill, R. (2010) 'Ricardo Bofill biography'. [Online] Available at: <http://www.ricardobofill.com> [Accessed 1 August].

Boutwell, S. (2008) 'Building Information Modeling and the adoption technologies'. *Extranet Evolution*. [Online] Available at: <http://www.extranetevolution.com/extranet_evolution/2008/06/bim-andsustainability.html> [Accessed June 2010].

Bowers, J. (1999) *Introduction to Two-Dimensional Design*. London: John Wiley and Sons.

Brandon, P. and Betts, M. (1995) *Integrated Construction Information*. London: E & FN Spon.

Bulletpoint (1996) *Creating a Change Culture – Not about Structures, but Winning Hearts and Minds*. New York: Wesley.

Burbidge, J. L. (1996) *Period Batch Control*. Oxford: Oxford University Press.

Cadazz (2010) 'CAD Software History'. [Online] Available at: www.cadazz.com/cad-software-history.htm [Accessed 1 January 2011].

Chau, K. W., Anson, M. and Zhang, J. P. (2005) '4D dynamic construction management and visualization software: 1. Development'. *Automation in Construction*, 14(4), 512–524.

Chin, S., Asce, M., Yoon, S., Choi, C., and Cho, C. (2008) 'RFID + 4D CAD for progress management of structural steel works in high-rise buildings'. *Journal of Computing in Civil Engineering* (April), 74–89.

Christiansson, P., Dalto, L., Skjaerbaek, J., Soubra, S. and Marache, M. (2002) 'Virtual environments for the AEC sector: the Divercity experience'. In *Proceedings of the International ECPPM Conference – eWork and eBusiness in Architecture, Engineering and Construction*, Portoroz, Slovenia.

Cleveland, A. B. (1989) 'Real-time animation of construction activities'. *Proceedings of Construction Congress I – Excellence in the Constructed Project*, pp. 238–243.

Cockburn, D. and Jennings, N. R. (1996) 'A distributed artificial intelligence system for industrial applications'. In O'Hare, G. M. P. and Jennings, N. R. (eds), *Foundations of Distributed Artificial Intelligence*. New York: Wiley, pp. 319–344.

Coles, B. C. and Reinschmidt, K. F. (1994) 'Computer-integrated construction: moving beyond standard computer-aided design to work in three and even four dimensions helps a project team plan construction, resolve conflicts and work more efficiently'. *Journal of Civil Engineering*, ASCE, 64(6).

Cooper, R. G. (1984) 'The performance impact of product innovation strategies'. *European Journal of Marketing*, 18(5), 223–229.

Cooper, R. G. (1990) 'Stage-Gate System: a new tool for managing new products'. *Business Horizons*, May–June, 44–54.

Cooper, R. G. (1993) *Winning at New Products: Accelerating the Process from Idea to Launch*. Reading, MA: Addison-Wesley.

Cooper, R. G. (1994) 'Third-generation new product processes'. *Journal of Product Innovation Management*, 10, 6–14.

Cooper, R. G. (1999) 'From experience: the invisible success factors in product innovation'. *Journal of Production Innovation Management*, 16, 115–133.

Cooper, R. G. and Kleinschmidt, E. J. (1987a) 'New products: what separates winners from losers?' *Product Innovation Management Journal*, 4, 169–184.

Cooper, R. G. and Kleinschmidt, E. J. (1987b) 'Success factors in product innovation'. *Industrial Marketing Management Journal*, 7, 9–21.

Cooper, R. G. and Kleinschmidt, E. J. (1995) 'Benchmarking the firm's critical success factors in new product development'. *Journal of Product Innovation Management*, 12, 374–391.

Cooper, R., Kagioglou, M., Aouad, G., Hinks, J., Sexton, M. and Sheath, D. (1998) 'Development of a generic design and construction process'. *European Conference on Product Data Technology*, BRE, pp. 205–214.

Coughlan, P. D. (1991) 'Differentiation and integration: the challenge of new product development. In Proceedings of the 5th Annual Conference of the British Academy of Management, June 28.

Crawford, C. M. (1977a) *New Products Management*. Burr Ridge, IL: Irwin.

Crawford, C. M. (1977b) 'Product development: today's most common mistakes'. *University of Michigan Business Review*, 6, 7–8.

Crawford, C. M. (1992) 'The hidden costs of accelerated product development'. *Journal of Product Innovation Management*, 9(3), 161–176.

Crawford, K. M. and Fox, J. F. (1990) 'Designing performance measurement systems for just-in-time operations'. *International Journal of Production Research*, 28(11), 2025–2036.

Cyon Research (2003) 'Building Information Model: a Look at Graphisoft's Virtual Building Concept'. Cyon Research Corporation.

Davenport, T. H. (1993) *Process Innovation – Reengineering Work through Information Technology*. Boston, MA: Harvard Business School Press.

Dawood, N., Sriprasert, E. and Mallasi, Z. (2003) 'Product and process integration for 4D visualisation at construction site level: a Uniclass-driven approach'. In Lee, A., Marshall-Ponting, A. J., Aouad, G., Wu, S., Koh, I., Fu, C., Cooper, R., Betts, M., Kagioglou, M. and Fischer, M. (2003) *Developing a Vision of nD-Enabled Construction*. Construct IT Report, Salford, pp. 64–68.

Dawood, N., Sriprasert, E., Mallasi, Z. and Hobbs, B. (2002), 'Development of an integrated information resource base for 4D/VR construction processes simulation'. *Automation in Construction*, 12 (2), 123–131.

Devinny, T. M. (1995) 'Significant issues for the future of product innovation'. *Journal of Product Innovation Management*, 12, 70–75.

Dinerstein, J., Egbert, P. K., Garis, H. and Dinerstein, N. (2004) 'Fast and learnable behavioral and cognitive modelling for virtual character animation'. *Computer Animation and Virtual World*, 15, 95–108.

Drogemuller, R. (2002) 'CSIRO & CRC-CI IFC Development Projects', ITM Meeting, Tokyo.

Durst, R. and Kabel, K. (2001) 'Cross-functional teams in a concurrent engineering environment – principles, model, and methods'. In Beyerlein, M. M. Johnson, D. A. and Beyerlein, S. T (eds), *Virtual Teams*. Oxford: JAI, pp. 167–214.

Eastman, C. (1999) *Building Product Models: Computer Environments Support Design and Construction*, Florida: CRC Press LLC.

Egan, J. (1998) *Rethinking Construction. Report from the Construction Task Force*, Department of the Environment, Transport and Regions, UK.

Elzinga, D. J., Horak, T., Chung-Yee, L. and Bruner, C. (1995) 'Business process management: survey and methodology'. *IEEE Transactions on Engineering Management*, 24(2), 119–128.

Emmerson, H. (1962) *Studies of Problems before the Construction Industries*. London: HMSO.

Evans, R. (1997) *Translations from Drawings to Building and other Essays*. London: Architectural Association, pp. 153–194.

Fenves, S. J. (1990) 'Integrated software environment for building design and construction'. Carnegie Mellon University. *Computer-aided Design*, 22(1), Jan/Feb.

Ferber, J. (1999) *Multi-Agent Systems: An Introduction to Distributed Artificial Intelligence*. Harlow, UK: Addison-Wesley.

Finkle, C. (2011) 'FireGL users for CAD, 3D design, film, and medical visualization. Rendering intent: to illustrate or to sell'. [Online] Available at: <http://fireuser.com/articles/solids_vs_surface_modeling_what_and_why_you_need_to> [Accessed February 2011].

Fischer, M. (1997) '4D modelling'. Proceedings of Global Construction IT Futures, Lake District, UK, April 1997, pp. 86–90.

Fischer, M. (2000) 'Construction planning and management using 3D & 4D CAD models'. Construction IT 2000, Sydney, Australia.

Fischer, M. (2000) 'Benefits of 4D models for facility owners and AEC service providers'. Construction Congress VI, ASCE, Orlando, Florida, February, 990–995.

Fischer, M. (2001) 'The frontier of virtual building'. Presentation given to the Workshop on Virtual Construction, organized by ENCORD, 26–27 November, Essen, Germany.

Fischer, M. and Liston, K. M. (2001) 'Wish list for 4D environments: a WDI R&D perspective'. Paperless Design Project Team at Walt Disney Imagineering. [Online] Available at: <http://www.stanford.edu/group/4D/issues/wishlist.shtml> [Accessed January 2011].

Freeman, S. (2009) 'FM issue: demystifying BIM, today's facility manager'. [Online] Available at: <http://www.todaysfacilitymanager.com/articles/fm-issue-demystifying-bim.php [Accessed June 2010].

Froese, T. (2008) 'The impact of emerging Information Technology on project management for construction'. *Automation in Construction*. [Online] Available at: <http://cedb.asce.org/cgi/WWWdisplay.cgi?147720> [Accessed May 2010].

Froese, T. and Paulson, B. (1994) 'OPIS: an object model-based project information system'. *Microcomputers in Civil Engineering*, 9, 13–28.

Gallimore, R. J., Jennings, N. R., Lamba, H. S., Mason, C. L. and Orestein, B. J. (1999) 'Co-operating agents for 3D scientific data interpretation'. *IEEE Transactions on Systems, Man and Cybernetics – Part C: Applications and Reviews*, 29(1).

Gillard, A., Counsell, J.A.M., and Littlewood, J.R. (2008) 'The Atlantic College case study – exploring the use of BIM for the sustainable design and maintenance of property'. The Construction and Building Research Conference of the Royal Institution of Chartered Surveyors COBRA.

Graphisoft, (2003) 'The Graphisoft virtual building: bridging the Building Information Model from concept into reality'. Graphisoft Whitepaper.

Graves, M. (1977) 'Conversation through drawing' *Architectural Design,* 6(77), 394–396.

Griffin, A. (1997) 'PDMA research on new product development practices: updating trends and benchmarking best practices'. *Journal of Product Innovation Management*, 14, 429–458.

GSA (2010) '3D-4D Building Information Modelling'. [Online] Available at: http://www.gsa.gov/portal/category/21062 [Accessed 1 September 2010].

Gunasekaran, A. and Love, P. E. D. (1998) 'Concurrent engineering: a multi-disciplinary approach for construction'. *Logistics Information Management*, 11(5), 295–300.

Gyles, R. (1992) *Royal Commission into Productivity in the New South Wales Building Industry*. Sydney: Government Printer.

Harvey, J. P. (1971) *The Master Builders – Architecture in the Middles Ages*. London: Thames & Hudson.

Heesom, D. and Mahdjoubi, L. (2002) 'Visualization development for the virtual construction site'. School of Engineering and the Built Environment, University of Wolverhampton.

Heesom, D. and Mahdjoubi, L. (2004) 'Trends of 4D CAD Applications for Construction Planning'. *Journal of Construction Management and Economics*, 22(2), 171–182.

Hill, T. J. (1992) 'Incorporating manufacturing perspectives in corporate strategy'. In Voss, C. A. (ed.) *Manufacturing Strategy*. Oxford: Chapman & Hall.

Hinks, J., Aouad, G., Cooper, R., Sheath, D., Kagioglou, M. and Sexton, M. (1997) 'IT and the design and construction process: a conceptual model of co-maturation'. *The International Journal of Construction*, July, 56–62.

HM Treasury (1998) *Innovating for the Future*. Department of Trade and Industry, London: HMSO.

Hoffman, L, R. (1979) *The Group Problem Solving Process: Studies of a Valance Model*. New York: Praeger.

Hohler, W. (2000) 'CAD/CAM Workshop: what is solid modeling?' [Online] Available at: <http://www.xmlcreate.com/NCGuide/Workshop/solids.html> [Accessed January 2011].

Howard, H.C. (1991) 'Linking design data with knowledge-based construction'. CIFE Spring Symposium, pp. 1–24.

Howell, D. (1999) 'Builders get the manufacturers in'. *Professional Engineer*, May, 24–25.

IDEF (2002) http://www.idef.com

Innovaya (2010) 'Innovaya Visual 4D Simulation'. [Online] Available at: http://www.innovaya.com/prod_vs.htm [Accessed 1 September 2010].

Jassawalla, A. R. and Sashittal, H. C. (1998) 'An examination of collaboration in high-technology new product development processes'. *Journal of Product Innovation Management*, 15, 237–254.

Jennings N. R., Faratin, P. J., Norman, T., O'Brien, P., Wiegand, M. E., Voudouris, C., Alty, J. L., Miah, T. and Mamdani, E. H. (1996) 'ADEPT: managing business processes using intelligent agents'. In Proceedings of BCS Expert Systems 96 Conference, Cambridge, UK, pp. 5–23.

Jongeling, R. and Olofsson, T. (2007), 'A method for planning of work-flow by combined use of location-based scheduling and 4D CAD'. *Automation in Construction*, 16 (2), 189–198.

Kagioglou, M., Cooper, R., Aouad, G., Hinks, J., Sexton, M. and Sheath, D. (1998a) 'Final report: generic design and construction process protocol'. The University of Salford, Salford.

Kagioglou, M., Cooper, R., Aouad, G., Hinks, J., Sexton, M. and Sheath, D. (1998b) 'A generic guide to the design and construction process protocol'. The University of Salford, Salford.

Kagioglou, M., Cooper, R., Aouad, G., Hinks, J., Sexton, M. and Sheath, D. (1998c) 'Cross-industry learning: the development of a generic design and construction process based on the Stage/Gate New Product Development Process found in the manufacturing industry'. In Proceedings of the Engineering Design Conference, Brunel, UK.

Kähkönen, K. and Leinonen, J. (2001a) 'Advanced communication technology as and enabler for improved construction practice'. Presentation given to Workshop on 4D Modelling: Experiences in UK and Overseas, organized by The Network on Information Standardization, Exchanges and Management in Construction, 17 October, Milton Keynes.

Kartam, N. (1994) 'ISICAD: interactive system for integrating CAD and computer-based construction systems'. *Microcomputers in Civil Engineering*, 9, 41–51.

Kartam, N. A. (1996) 'Making effective use of construction lessons learned in project life cycle'. *Journal of Construction Engineering and Management*, March, 14–21.

Katzenbach, J. (1996) *Real Change Leaders*. London: Nicholas Brearley.

Khurana, A. and Rosenthal, S. R. (1998) 'Towards holistic "front ends" in new product development'. *Journal of Product Innovation and Management*, 15, 57–74.

Kmethy, G. (2008) 'ArchiCAD versions – ArchicadWiki' FrontPage – ArchicadWiki. [Online] Available at: <http://www.archicadwiki.com/ArchiCAD%20versions> [Accessed January 2011].

Koo, B. and Fischer, M. (2000) 'Feasibility study of 4D CAD in commercial construction'. *Journal of Construction Engineering and Management*, ASCE, 126(4), 251–260.

Koskela, L. (1992) 'Application of the new production philosophy to construction'. Technical report no. 72. Centre for Integrated Facility Engineering, Stanford University.

Kuczmarski, T. D. (1992) *Managing New Products: The Power of Innovation*. Englewood Cliffs, NJ: Prentice Hall.

Kumaraswamy, M. M. and Chan, D. W. M. (1998) 'Contributors to construction delays'. *Construction Management and Economics Journal,* 16(1), 17–29.

Kunz, J., Fischer, M., Haymaker, J., and Levitt, R. (2002) 'Integrated and automated project processes in civil engineering: experiences of the Centre for Integrated Facility Engineering at Stanford University'. Computing in Civil Engineering Proceedings, ASCE, Reston, VA, 96–105, January 2002.

Kymell, W. (2008) *Building Information Modeling: Planning and Managing Construction Projects with 4D CAD and Simulations.* New York: McGraw-Hill.

Latham, M. (1994) *Constructing the Team: Joint Review of Procurement and Contractual Arrangements in the UK Construction Industry.* Department of the Environment, London: HMSO.

Lawson, B. R. (1990) *How Designers Think* (second edition). Oxford: Butterworth Architecture.

Lawson, B. R. (1994) *Design in Mind.* Oxford: Butterworth Architecture.

Lawson, B. R. (2004) 'Oracles, draughtsman and agents: the nature of knowledge and creativity in design and the role of IT'. In Proceedings of the International Conference on Construction Information Technology (INCITE) conference, Malaysia, pp. 9–16.

Lee, A., Betts, M., Aouad, G., Cooper, R., Wu, S. and Underwood, J. (2002b) 'Developing a vision for an nD modelling tool' (key note speech). In proceedings of CIB W78 Conference – Distributing Knowledge in Building (CIB w78), Denmark, 141–148.

Lee, A., Marshall-Ponting, A.J., Aouad, G., Wu, S., Koh, I., Fu, C., Cooper, R., Betts, M., Kagioglou, M. and Fischer, M. (2003) *Developing a Vision of nD-Enabled Construction,* Construct IT Report, ISBN: 1–900491–92–3.

Lee, A., Wu, S., Aouad, G. and Fu, C. (2002a) 'Towards nD modelling'. Submitted to the European Conference on Information and Communication Technology Advances and Innovation in the Knowledge Society. E-sm@art, Salford.

Lee, A., Wu, S., Aouad, G., Cooper, R., Tah, J. H. M. and Marshall-Ponting, A. (2005) 'A roadmap for nD-enabled construction'. *RICS Paper Series,* 6(2), May.

Lee, A., Wu, S., Marshall-Ponting, A.J., Aouad, G., Abbott, C. A., Cooper, R. and Tah, J. (2005) *nD Modelling Roadmap: A Vision of the Future of Construction.* Centre for Construction Innovation, Salford.

Lefebvre, H. (1991) *The Production of Space* (trans. D. Nicholson-Smith). Oxford: Blackwell.

Leibich, L., Wix, J., Forester, J. and Qi, Z., (2002) 'Speeding-up the building plan approval – the Singapore e-Plan checking project offers automatic plan checking based on IFC'. The International Conference of ECPPM 2002 – eWork and eBusiness in Architecture, Engineering and Construction, Portoroz, Slovenia, 2002.

Lenerd, O. (2010) 'Revit BIM Experience Award to ONL'. Press Release.

Li, H. and Love, P. E. D. (1998) 'Developing a theory of construction problem solving'. *Construction Management and Economics,* 16, 721–727.

Lingle, J. H. and Schiemann, W. A. (1996) 'From balanced scorecard to strategy gauge: is measurement worth it?' *Management Review,* March, 56–52.

Liston, K., Fischer, M. and Winograd, T. (2001) 'Focused sharing of information for multidisciplinary decision making by project teams'. *ITcon,* 6, 69–82. [Online] Available at: http://www.itcon.org/2001/6.

Lundgren (2002) Process. Unpublished proposal.

Ma, Z., Shen, Q. and Zhang, J. (2005) 'Application of 4D for dynamic site layout and management of construction projects'. *Automation in Construction,* 14(3), 369–381.

Madsen, J. (2008) Build Smarter, Faster, and Cheaper with BIM, Buildings, [Online] http://www.buildings.com/ArticleDetails/tabid/3321/ArticleID/6149/Default.aspx [Accessed May 2010]

Martin, J. and Odell, J. (1992) *Object Oriented Analysis and Design.* Englewood Cliffs, NJ: Prentice Hall.

MbDesign (2010) 'CAD history'. [Online] Available at: http://mbinfo.mbdesign.net/CAD-History.htm [Accessed 1 January 2011].

McGarth, M. E. (1996) *Setting the Pace in Product Development.* Boston, MA: Butterworth-Heinemann.

McGraw Hill Construction (2011) 'Building products in BIM'. McGraw Hill Construction. [Online] Available at: <http://continuingeducation.construction.com/article.php?L=251&C=775> [Accessed Jun 2010].

McKinney, K. and Fischer, M. (1998) 'Generating, evaluating and visualizing construction schedules with CAD tools'. *Automation in Construction*, 7(6), 433–447.

Mitchel, J. (2005) Sydney Opera House – FM Exemplar Project, Report Number: 2005–001-C-3, Open Specification for BIM: Sydney Opera House Case Study, QUT Digital Repository.

Mitchell, J., and Schevers, H. (2006) Building Information Modelling for FM using IFC.

MOB (Modeles Objet Batiment) (1994) Rapport Final, Appel d'offres du Plan Construction et Architecture, Programme Communication/Construction.

Mohsini, R. A. and Davidon, C. H. (1992) 'Detriments of performance in the traditional building process'. *Journal of Construction Management and Economics*, 10, 343–359.

Moran, J. W. and Brightman, B. K. (1998) 'Effective management of healthcare change'. *The TQM Magazine*, 10(1), 27–29.

Musse, S. R. and Thalmann, D. (2001) 'Hierarchical model for real time simulation of virtual human crowds'. *IEEE Transactions on Visualization and Computer Graphics*, 7, 152–164.

Newnham, L. N., Anumba, C. J., and Ugwu, O. O. (1999) 'Negotiation in a multi-agent system for the collaborative design of light industrial buildings'. Technical Report No. ADLIB/02, Loughborough University, UK, October 1999.

NIST (2004) 'Cost analysis of inadequate interoperability in the U.S. capital facilities industry'. [Online] Available at: <http://fire.nist.gov/bfrlpubs/build04/art022.html> [Accessed May 2010].

Oakland, J.S. (1995) *Total Quality Management: The Route to Improving Performance.* 2nd ed. Boston, MA: Butterworth Heinemann Ltd.

Okeil, A. (2010) 'Hybrid design environments: non-immersive architectural design'. *Journal of Information Technology in Construction*, 15, 202–216.

Parunak, H. D. (1999) 'Experiences and issues in the development and deployment of industrial agent-based systems'. In the Proceedings of the 4th International Conference on the Practical Application of Intelligent Agents and Multi-Agent Technology (PAAM99), pp. 3–9.

Park, J., Kim, B., Kim, C., and Kim, H. (2011). '3D/4D CAD applicability for life-cycle facility management'. *Computer* (April), 129–138.

Penn, A. and Turner, A. (2002) 'Space syntax based agent simulation'. In Schreckenberg, M, and Sharma, S. (eds) *Pedestrian and Evacuation Dynamics.* Berlin: Springer-Verlag, pp. 99–114.

Peppard, J. and Rowland, P. (1995) *The Essence of Business Process Re-engineering.* Reading, MA: Prentice Hall.

Plossl, K. R. (1987) *Engineering for the Control of Manufacturing.* Englewood Cliffs, NJ: Prentice-Hall.

Powell, J. (1995) 'Virtual reality and rapid prototyping for engineering'. Proceedings of the Information Technology Awareness Workshop, University of Salford.

Rasdorf, N. J. and Abudayyeh, O. (1992) 'NIAM conceptual database design in construction management'. *Journal of Computing in Civil Engineering*, 6(1), 41–62.

Rezgui, Y. A., Brown, G., Cooper, R., Aouad, A., Kirkham, J., and Brandon, P. (1996) 'An integrated framework for evolving construction models'. *The International Journal of Construction IT*, 4(1), 47–60.

RIBA (1997) *RIBA Plan of Work for the Design Team Operation*. 4th edn. London: Royal Institute of British Architects Publications.

Riedel, J. C. K. H. and Pawar, K. S. (1997) 'The consideration of production aspects during product design stages'. *Integrated Manufacturing Systems*, 8(4), 208–214.

Rischmoller, L. and Matamala, R. (2003) 'Reflections about nD modelling and computer advanced visualisation tools (CAVT)'. In Lee, A., Marshall-Ponting, A. J., Aouad, G., Wu, S., Koh, I., Fu, C., Cooper, R., Betts, M., Kagioglou, M. and Fischer, M. (2003) *Developing a Vision of nD-Enabled Construction*. Construct IT Report, Salford, 92–94.

Rischmoller, L., Fisher, M., Fox, R. and Alarcon, L. (2000) '4D Planning and scheduling (4D-PS): grounding construction IT research in industry practice'. In Proceedings of CIB W78 conference on Construction Information Technology: Taking the construction industry into the 21st century, June, Iceland.

Robbins, E. (1997) *Why Designers Draw*. Cambridge, MA: MIT Press.

Schon, D. A. (1983) *The Reflective Practitioner: How Professionals Think in Action*. London: Temple Smith.

Schonberger, R. J. (1982) *Japanese Manufacturing Techniques: Nine Hidden Lessons in Simplicity*. New York: Free Press.

Shapiro, V. (2001) *Geometric Models and Representations: Handbook of Computer Aided Geometric Design*. Amsterdam: Elsevier Science Publishers, pp. 473–518. [Online] Available at: <http://sal-cnc.me.wisc.edu/index.php?option=com_remository&Itemid=143&func=fil> [Accessed January 2011].

Sheath, D. M., Woolley, H., Cooper, R., Hinks, J. and Aouad, G. (1996) 'A process for change: the development of a generic design and construction process protocol for the UK construction industry'. In Proceedings of the CIT Conference, Institute of Civil Engineers, April, Sydney, Australia.

Sidwell, A. C. (1990) 'Project management: dynamics and performance'. *Construction Management and Economics*, 8, 159–178.

Smith, V. S. (2009) 'CAD architectural 3D design = 3D CAD drawings for building construction'. Ezine Articles. [Online] Available at: <http://ezinearticles.com/?Architectural-3D-Design—-3D-CAD Drawings-For-Building-Construction&id=2476892> [Accessed February 2011].

Sower, V. E., Motwani, J. and Savoie, M. J. (1997) 'Classics in production and operations management'. *International Journal of Operations and Production Management*, 17(1), 15–28.

Stern, R. (1977) 'Drawing: Towards a more modern architecture'. *Architectural Design,* 6(77), 382–390.

Takeuchi, H.and Nonaka, I. (1986) *The New Product Development Game*. Cambridge, MA: Harvard Business Press.

Tanyer, A.M. and Aouad, G. (2005) 'Moving beyond the fourth dimension with an IFC-based single project database'. *Automation in Construction,* 14 (1), 15–32.

THOCP (2010) 'The History of Computing Project'. [Online] Available at: http://thocp.net/ [Accessed 1 January 2011].

Tidd, J., Bessant, J. and Pavitt, K. (1997) *Managing Innovation*. Chichester: John Wiley & Sons.

Tu, X. and Terzopoulos, D. (1994) 'Artificial fishes: physics, locomotion, perception, behavior'. In Proceedings of ACM SIGGRAPH, pp. 43–50.

United Nations (1959) *Government Policies and the Cost of Building*. Geneva: ECE.

Vonderembse, M. A. and White, G. P. (1996) *Operations Management: Concepts, Methods and Strategies*. New York: West Publishing.

Wang, H.J., Zhang, J.P., Chau, K.W., Anson, M. (2004) '4D Dynamic management for construction planning and resource utilization'. *Automation in Construction*, 13(5), 575–589.

Watson, A. and Crowley, A. (1995) 'CIMSteel integration standard'. In Scherer, R. J. (ed.), *Product and Process Modelling in the Building Industry*. Rotterdam: A.A. Balkema, pp. 491–493.

Webb, R. M. (2000) '4D-CAD: construction industry perspective'. In Walsh, K. D. (ed.) *Construction Congress VI*. Orlando, FL: ASCE, pp. 1042–50.

White, A. (1996) *Continuous Quality Improvement: A hands-on Guide to setting up and sustaining a Cost Effective Quality Programme*. London: Piatkus.

Yessios, C. I. (2004) 'Are we forgetting design AEC bytes viewpoint'. [Online] Available at: <http://www.aecbytes.com/viewpoint/2004/issue_10.html> [Accessed June 2010].

Yu, K., Froese, T. and Grobler, F. (2000) 'A development framework for data models for computer-integrated facilities management'. *Automation in Construction,* 9, 145–167.

Zairi, M. (1997) 'Business process management: a boundary-less approach to modern competitiveness'. *Business Process Management*, 3(1), 64–80.

Zhang, J., Anson, M. and Wang, Q. (2000) 'A new 4D management approach to construction planning and site space utilization'. 8th International Conference on Computing in Civil and Building Engineering, ASCE, Stanford University California, USA, 14–17 August.

Index

In this index numbers are filed as words. For example 2D drafting is filed under 'two'; 4D modelling under 'four'.

abstraction, information modelling 26
AEC/FM industry: advantages of 3D modelling 53–4; and BIM 71, 72, 74–81, 91; and globalisation 112; 3D modelling for 52–3
analysis: and BIM 76–8, 82, 90–1; construction schedules and 4D CAD 94
analytical hierarchy process (AHP) 111
analytical reductionism/process decomposition 21–2
ArchiCAD 9; and BIM 72, 73, 85, 87, 88; rendering software 65; vertical toolbox 89
architectural drafting, training in 36
arraying, and editing lines 39
AutoCAD 7, 8, 9; extrusion tools 60, 61; and layering 38; surface models 66; 3D model generation 51; and 3D volumes 55; viewports 63
AutoCAD 3D 55
AutoCAD LT 41, 43, 44, 45, 46, 47
AutoCAD LT, viewports 62, 63
Autodesk Architectural Desktop (ADT) 73
Autodesk Navisworks, TimeLiner tool 100
Autodesk Revit 73, 100; horizontal ribbon 89; 3D modelling 100, 101
avatars, intelligent 114–15

behavioural model, and virtual characters 115
Bentley Navigator 99
Bentley Systems 3, 73
BIM 8–9, 71–2, 74; and accuracy 82, 83; adoption rate 72; advantages of 81–3; and aecXML 74; and AEC/FM industry 71, 72, 74–81, 91; align tool 86; and architects 75–8; and bid preparation 79; and building code compliance 77; and building maintenance 80; and building procurement process 75; and building records 80; and buildings management 79; and clash detection 77; code compliance checks 77, 86, 91; and communication 74; and computer speed 72; and construction bidding 79; and construction planning 81; and construction specifications 78; for contractors 78–9; copy tool 88; creating a BIM model 85–91; database 84; and data extraction/analysis 82; data requirement 83; for designers 75–8; and design flexibility 82; and design productivity 81; and design review 75–6; and detailing 76; disadvantages of 83; and documentation 76, 81, 82; door schedule 86, 90; and energy consumption models 80; and engineers 75–8; and environmental performance of buildings 83; and estimation 76; example model 87–8, 89; and facility managers 79–81; and fire analysis 76–7; and 4D CAD 92, 94; history of 72–3; and Industry Foundation Classes (IFC) 74; and interoperability 81–2, 90; and lifecycle costs of buildings 83; and management of buildings 79; and application of materials 90; and material takeoffs 82; and mirror tool 86, 88; and model editing 86; modelling principles 83–4; model setup 84; and model validation/visualisation 90–1; and modification 81, 82; modification tools 86; move tool 86; occupancy tracking, and BIM

81; and offsite manufacturing 79; and
onsite construction 78–9; Open Standards
Driven 84; and parametric modelling 84;
and planning/scheduling 78; and pre-cast
concrete 79; and problem identification 81;
and program validation 85; purpose of
73–4; and quantification/estimation 76; and
rendering 76, 90; and REVIT Architecture
78, 84, 85, 86, 88; and roof modelling 86;
and room finish schedule 90; rotate tool 88;
and schedule creation 88–90; and
simulation 80–1; software 72–3, 77, 83;
and space layout/management 80, 85; and
steel fabrication 79; and sustainability
77–8; and textures, application 90; and 3D
design 75; and 3D views 82; and time
element 78; and trade coordination 78;
virtual world potential of 74; and
visualisation 75–6; and whole-life costs of
buildings 83; and window schedules 90
'brick wall' approach, design and
construction 14
Building Information Modelling see BIM
building management systems 79
building process, and United Nations 12–13
building process, traditional 13
BuildingSMART 74

CAD: defined 1–3; history of 3–9; modern
applications 8–9; and parts/products 2;
and PCs 7; and Plan of Works 15–16;
software 2; systems 3–6, 7; technology
9–10
CAD CAM 3
CADD 2
CADKEY 3
Cambridge University, Computing
Laboratory 52
CAM systems, and solid modelling 57
CATIA 9, 52
Citroen 52
clash detection 52, 77
clients/users, translating design thoughts 2
cognitive model, and virtual characters 115
communication 74; design/3D models 53;
and 4D CAD 95; and nD model 109
construction group, traditional building
process 13
construction industry: performance of 11,
12; and product modelling 25–6; surface
modelling 58
construction schedules: and 4D CAD 96,
98–9; Naviswork Timeliner 103

cost, of CAD packages 9–10
cross-functional process models 20–1
customer satisfaction, and cross-functional
process models 20

design: criteria for successful 108; three-
dimensional view 14
design changes: and stakeholders 107–8;
traditional building process 13–14
design group, traditional building process 13
drafting 31
drafting departments, and CAD 8
dynamic mathmatical modelling, and CAD 2

efficiencies, CAD drafting 33–4
encapsulation, information modelling 26
Entity-Based CAD 7–8
EPSRC (Engineering and Physical Sciences
Research Council), Platform grant/nD
research 110–11
error reduction: and 4D CAD 97; and 3D
prototypes 54
extrusion: 3D models 55; of 2D shape to
create 3D object 60–1
extrusion tools 51

flow charts 18–20
form, and design 2
4D CAD 92; and activity duration
97;advantages of 96; and BIM 92, 94;
challenges to overcome 104; and
collaborative working 94; and
construction management 94; and
construction phase 96; and construction
planning 97; and construction projects 92;
and design stage 95–6; and feasibility stage
95; historical development of 94; and
integrated project databases 95; limitations
of 97; and location-based schedule method
94; modelling process 97–104; practical
example of 100–4; in practice 95–6; and
pre-construction phase 95–6; and project
lifecycle 95–6; and resources management
94; and safety management 96; and site
logistics/layout 94; software 99–100; and
3D CAD model/BIM model 98; and waste
costs 94
freeform/NURB surface modelling 59
function models 23

Gantt charts 94, 97
General Services Administration (GSA)
Public Buildings Service (PBS) 74

geometric constraint, and solid modelling
    programs 58
globalisation, and AEC/FM industry 112
graphical representation, and BIM 76
Graphisoft 72, 73

hardware/operating system (OS)
    technologies 9

IDEF 22–4
IDEF0 22–4; decomposition structure 24
IDEF1 22
IDEF1X 22
IDEF2 22
IFCs see Industry Foundation Classes
IGES (Initial Graphics Exchange
    Specification) 7, 28
Industry Foundation Classes (IFC): and
    interoperability 28, 81, 109, 112
information modelling: abstraction 26;
    encapsulation 26; inheritance 26, 27;
    polymorphism 26
inheritance, information modelling 26, 27
Innovaya Visual 4D simulation 100
Integrated DEFinition language 0 see IDEF0
interface, CAD applications 10
International Alliance for Interoperability
    (IAI) 109
International Organisation for
    Standardisation 28
Interoperability;:and Industry Foundation
    Classes 28, 81, 109, 112; and nD
    modelling 109; software 33; 3D packages
    54–5
ISO 10303 28
IT, future of in construction 112

lighting, and rendering programs 65

MAS see multi–agent system (MAS)
Massachusetts Institute of Technology (MIT)
    3, 52
mirroring, and editing lines 39
model checking programs 90–1
modelling: solid 57–8; surface 58–9
modelling tools, active 89
modify ribbon, REVIT Architecture 88
MS Project 2007 100; construction schedule
    101; and Naviswork Timeliner 103
multi-agent system (MAS): and conflict
    resolution of modern building design 114;
    2G:nD modelling environment 114
multi-criterion decision analysis (MCDA) 111

National Institute of Standards and
    Technology, and interoperability 81–2
Naviswork 77, 100, 102, 103, 104, 105
nD knowledge base 111
nD modelling 107–15; and documentation
    109; explained 109; future of 112–15;
    and ICT 112; and project cost
    information 94; research development
    110–11
nD modelling research framework 113,
    114
nD Modelling Roadmap: A Vision for
    nD-Enabled Construction 112
nD modelling tool, system architecture of
    111
nD prototype tool 114
Nonuniform Rational B-Spline modelling see
    NURBS
NURBS surface modelling 59;
    modifying/shaping objects 63

object-oriented CAD 8, 109
object-oriented models 26; using IDEF0 27
object-oriented paradigm, and perspectives
    27
over-the-brick-wall-effect 108

parts, and CAD 2
PCs, and CAD software 7 see also computer
    speed
PDES (Product Data Exchange Specification)
    7, 28
perspectives, and object-oriented paradigm
    27
photorealistic rendering 65
Podium 65, 69
polyline, trimming of 39
polymorphism, information modelling 26
procedure, defined 22
process, a 21–2; defined 16–18; flow charts
    18–20
process levels 21
process modelling 11, 12, 17–25, 28
Product Data Technology (PDT) 28
production processes, and process/project
    modelling 29
product modelling 11, 25–7, 28
product orientation, and reusability 26–7
product-oriented paradigm models 26
products, and CAD 2
project modelling, need for improved 14
ProjectWise Schedule Simulation 99–100
push/pull tool, surface modelling 66, 67

regular surface modelling 58
rendering: photorealistic 51, 65; 3D
   architectural 51; 3D modelling 64–5;
   wireframe 65–6
reusability, and product orientation 26–7
REVIT Architecture *see* Autodesk Revit
Royal Institute of British Architects (RIBA)
   Plan of Work 14–16

SET (Standard d'Echange et de Transfert)
   28
simulation, using virtual characters
   114–15
Sketchpad 3, 32
Sketchup: extrusion tools 60, 61;
   rendering software 65, 69; surface
   models 66; and 3D modelling 54, 55;
   and 3D volumes 55
Sketchup 7 66
soft clashes 77
software: 4D modelling 99–100;
   interoperability 33; model checking
   programs 90–1; rendering 51, 64–5;
   surface modelling 59; 3D programs 54–5;
   3D/viewports 62–3 *see also* names of
   packages
Solibri Model Checker 90

stakeholders, building 107
Standard Classification Methods (Uniclass)
   94
Standard for Exchange of Product data
   (STEP) 28
surface modelling 58–9; and computer speed
   60; practical example of 66
surface modelling software 59
systems, and process/project modelling 29

3D 'dumb' solids 7
3D modelling 51; advantages of 53–4; and
   competitiveness 53; and construction stage
   foresight 54; and coordinate systems 57;
   and detail 56; and design communication
   53; disadvantages of 54–5; and
   exaggeration 56; and functionality 57;
   history of 51–2; key features of 53; and
   light 56; and marketing 53;
   modifying/shaping 3D objects 63–4; orbit
   tool 63; pan tools 63; practical examples
   of 66–9; principles of 55–9; prototyping
   attributes of 53; purpose of 52–3; and
   realism 55; and scale/proportion 55; and
   shade 56; slicing 64; sphere volumes 55;

3D mesh 55–6; texturing/surface
   treatments 56, 64–5; nd volumes 55, 56;
   and walkthroughs 53; zoom tools 63
3D models, creation of 60–6
3D objects, creating from 2D objects 60–1
3D parametric solid modelling 8
3D software programs, viewports 62–3
3D space, navigating in 61–3
3D Studio Max, viewports 63
3D to nD modelling project 112
3D wireframe 7
three-dimensional view, design 14
time dependent clashes 77
training: in 3D modelling 54; 2D manual
   drawing/drafting 36
trimming, and editing lines 39
2D CAD drafting 31; advantages of 33–5;
   annotation in 41; circles/ovals 44–5; and
   competitive pricing 34; creating arcs 43–4;
   creating a new drawing 37; and
   dimensioning 41; disadvantages of 36;
   drawing grid 37; drawing limits 37–8;
   drawing units 37; hatched wall with door
   47–8; hatching, 41, 45–6; history of 31–2;
   layering 38; lines/shapes 38–42; object
   properties in 40–1; ; practical examples
   41–8; principles of 36–41; purpose of
   32–3; and scaling 40; snap 37;
2D systems 7
2G:nD, and virtual characters 114–15
2G:nD modelling environment, multi-agent
   system (MAS) 114
2G:nD project 112
2G:nD tool 113

UML (Unified Modelling Language) 26
United Nations, and building process
   12–13
University of Salford 107, 110, 112
UNIX workstations 7
US Air Force: Integrated Information
   Support System program 22; Program for
   Integrated Computer Aided Manufacturing
   (ICAM) 22
users/clients, translating design thoughts 2

Virtual Building Solution 72
virtual characters, intelligent 114–15
virtual-reality interactive models 53

what-if analysis: and 4D CAD 96; nD
   modelling 110; and 2G:nD 114
wireframe rendering 65–6